About This Book

If you have ever asked, "Why am I here?" or "What are my lessons?" then this book will help you to objectively facilitate the answers from an astrological point of view.

If you are a professional counselor whose clients come to you with these types of questions, then this book will serve as an aid in helping you to objectively answer these questions on behalf of your clients.

Beyond presenting key principles and ideas about the nature of the evolutionary journey of the Soul, this book supplies practical, concise and specific astrological methods and techniques that pinpoint the answers to the above questions. Jeff Green has spent the past ten years developing the methods and techniques given in this book and he has worked with thousands of clients who have asked these very questions. The methods given here are the results of those ten years.

The reader who studies this material carefully and then applies it to his or her own chart, will discover an objective vehicle to uncover the essence of his or her own state of being. The understanding that this promotes can help the individual cooperate with, instead of resist, the evolutionary and karmic lessons of his or her life.

Among other things, the author shatters the often repeated accusation that when astrologers address the spiritual/metaphysical aspect of life experience, astrology becomes foggy and irrelevant. Jeff Green makes the Grand Sweep. Individual destiny reflects the larger universal design. His deft interpretations simultaneously touch the Soul while clearly delineating the practical issues of one's personal journey on this planet. Yes, this is a monolithic work on Pluto, but it is so much more. Surely Jeff Green is here as an escort through the evolutionary leap in progress.

— **Diana Stone**

This book could be titled, "Brave New Book" — it will impact the right brain and should thus be absorbed, rather than read. Jeff includes whole brain techniques, however, and though it is a thoroughly metaphysical book, it is an imminently practical handbook for all astrological disciplines. As the only book of this magnitude on Pluto alone, it fills a great gap in our astrological libraries; as a result, it is a must. Very easy to read, it has a conversational tone, and the depth of insight into the Plutonian realms becomes readily available and useful. The manifestations of Shiva come to life in the application of the symbol of Pluto — god of the seed of incarnation. Progress is preceded by annihilation preceded by progress — thus we evolve toward the unknown perfection out of the perfect path. Absorb this book.

— **Erin Sullivan-Seale**

About the Author

Jeff Green has been a full-time, practicing professional astrologer for the past fourteen years in Seattle, Washington. Besides having an extremely busy practice, he has lectured for many organizations throughout the United States and Canada, including being a faculty member at most major conferences such as the Third World Congress of Astrology in Zurich, Switzerland, AFA, UAC, ISAR, WRAC, NORWAC, SWAC, Canada's Regional Conferences in Toronto and Edmonton, and the Canadian National Conference in Vancouver. He has had many articles published in American, Canadian, and European astrological magazines, and is the author of the best-selling book *Pluto: The Evolutionary Journey of the Soul*. Jeff spent two years in association with a Vedantic temple, and his formal education is in psychology and philosophy. Jeff is also a Vietnam veteran, having served with the U.S. Marines.

To Write to the Author

If you wish to contact the author or would like more information about this book, please write to the author in care of Llewellyn Worldwide, and we will forward you request. Both the author and publisher appreciate hearing from you and learning of your enjoyment of this book and how it has helped you. Llewellyn Worldwide cannot guarantee that every letter written to the author can be answered, but all will be forwarded. Please write to:

Jeff Green
c/o Llewellyn Worldwide
P.O. Box 64383-296, St. Paul, MN 55164-0383, U.S.A.

Please enclose a self-addressed, stamped envelope for reply, or $1.00 to cover costs.
If outside the U.S.A., enclose international postal reply coupon.

Free Catalog from Llewellyn

For more than 90 years Llewellyn has brought its readers knowledge in the fields of metaphysics and human potential. Learn about the newest books in spiritual guidance, natural healing, astrology, occult philosophy and more. Enjoy book reviews, new age articles, a calendar of events, plus current advertised products and services. To get your free copy of *Llewellyn's New Worlds of Mind and Spirit*, send your name and address to:

Llewellyn's New Worlds of Mind and Spirit
P.O. Box 64383-296, St. Paul, MN 55164-0383, U.S.A.

THE LLEWELLYN MODERN ASTROLOGY LIBRARY

Books for the *Leading Edge* of practical and applied astrology as we move toward the culmination of the 20th century.

This is not speculative astrology, nor astrology so esoteric as to have little practical application in meeting the needs of people in these critical times. Yet, these books go far beyond the meaning of "practicality" as seen prior to the 1980's. Our needs are spiritual as well as mundane, planetary as well as particular, evolutionary as well as progessive. Astrology grows with the times, and our times make heavy demands upon Intelligence and Wisdom.

The authors are all professional astrologers drawing from their own practice and knowledge of historical persons and events, demonstrating proof of their conclusions with the horoscopes of real people in real situations.

Modern Astrology relates the individual person to the Universe in which he/she lives, not as a passive victim of alien forces but as an active participant in an environment expanded to the breadth, *and depth,* of the Cosmos. We are not alone, and our responsibilities are infinite.

The horoscope is both a measure, and a guide, to personal movement—seeing every act undertaken, every decision made, every event, as *time dynamic:* with effects that move through the many dimensions of space and levels of consciousness in fulfillment of Will and Purpose. Every act becomes an act of Will, for we extend our awareness to consequences reaching to the ends of time and space.

This is astrology supremely important to this unique period in human history, when Pluto transits through Scorpio and Neptune through Capricorn, and the books in this series are intended to provide insight into the critical needs and the critical decisions that must be made.

These books, too, are "active agents", bringing to the reader knowledge which will liberate the higher forces inside each person to the end that we may fulfill that for which we were intended.

Carl Llewellyn Weschcke

OTHER BOOKS BY JEFF GREEN

Uranus: Freedom From The Known
Pluto: The Evolutionary Journey of the Soul, Volume Two

Forthcoming:

Psychodynamics of the Moon
Key Planetary Pairs and Phases
   (This is a book that I was working on with the late Robert Jansky.
   We had planned to co-author this book. He was going to rewrite
   his book *How to Interpret Aspects* and weld it to the book on
   Key Planetary Pairs and Phases that I was writing as my half of
   the project. We were about half finished when he left his body.
   This project has been on hold ever since.)
Planetary Nodes: Their Meaning and Application In The Birth Chart

PLUTO

THE EVOLUTIONARY JOURNEY OF THE SOUL

Volume I

Jeff Green

1996
Llewellyn Publications
St. Paul, Minnesota 55164-0383, U.S.A.

FIRST EDITION
Twelfth Printing, 1996

Cover Graphic: Castor & Pollux Studio

Library of Congress Cataloging-in-Publication Data
Green, Jeff, 1946-
    Pluto, the evolutionary journey of the soul.

    1. Astrology. 2. Pluto (Planet)—Miscellanea.
3. Horoscopes. I. Title.
BF1742.2.P4G74 1985   133.5'3   85-45290
ISBN 0-87542-296-9

Llewellyn Publications
A Division of Llewellyn Worldwide, Ltd.
P.O. 64383, St. Paul, MN 55164-0383

Dedicated to Paramahansa Yogananda, whose love, care, forgiveness, and guidance inspired this work . . . and to all of you who kept prodding me to get this book done. You know who you are. God Bless you all.

A special dedication to Mr. Noel Tyl whose special attention and relentless efforts allowed this book to be published in the manner that it has.

ACKNOWLEDGEMENTS

I wish to thank my friend, lover, and wife Laurie for the sustained belief in my work, especially in my moments of quiet desperation and futility. I wish to thank her for the deep personal sacrifices she has made that have allowed me to do this work.

I wish to thank Lynn Weyand for the timely financial and spiritual assistance she has given to allow this work to take form, Lucille Baker and Thia Bell for their tireless efforts in the initial editing and typing of this work, Bret Bowman and Karuna Baum for their historical research efforts, and Lyn Weyand, Laurie Burnett-Green, June Novak, Laura Gerking, and Margaret Nalbandian for their valued comments after reading the manuscript. Laura also spent a great deal of time doing the calligraphy for the charts in this book. My deepest gratitude goes to Sandy LaForest for her final editing of this book.

I also wish to thank Eileen Nauman for teaching me how to write in a succinct fashion, and for helping me to expose the ideas in this book at various conferences by helping to create the openings. I wish to thank the late Robert Jansky for helping me to believe in myself, Joan McEvers for her encouragement, inspiration, and gentle chidings to get me to write, Margaret and Noubar Nalbandian for their sustained support over all the years I have been involved with astrology, and Saulis Pempe and Theresa Maybee of Castor and Pollux Studios in Seattle for their dedicated efforts in creating the original photographic artwork that has led to the wonderful cover for this book.

My special thanks to Donna Cunningham, Joan McEvers, Erin Sullivan-Seale, and Diana Stone for taking the time to review the manuscript, to Alan Oken for his inspired Forward, and Noel Tyl for his wonderful Preface. A special thanks also to Terry Buske at Llewellyn for her superb editing and Carol Maki for working on her weekends to get this book done.

And last, but by no means least, to all of you whom I have run into over the years at various workshops, lectures, and conventions that have inspired me to keep on truckin'. God Bless you all.

"A child is born on that day and at that hour when the celestial rays are in mathematical harmony with his individual karma. His horoscope is a challenging portrait, revealing his unalterable past and its probable future result. But the natal chart can be rightly interpreted only by men of intuitive wisdom: these are few."

Swami Sri Yukteswar

CONTENTS

PREFACE

BY NOEL TYL

"Where is Billy?" — A good test question for an infant in the midst of self-discovery. But just where *is* Billy? Is he that which displaces a specified amount of space and/or a specified portion of time? Or is he simply "right here", to be conveyed with some kind of impersonal gesture?

Psychological tests of early concepts of identity suggest that the majority of Billys (and Janes, of course) will answer this existential question by pointing to their stomachs, their midsections.

As you read these paragraphs, ask yourself where *you* are. Say it out loud. What are your feelings? It's an awkward moment that can't be easily dismissed.

So often in life we reply to circumstance with the phrase/thought "Who ME?" Say this to yourself right now, loud and strong.

Another strange feeling. Either you did make a gesture to accompany the two words, or you felt inclined to do so (a twitch in the wrist perhaps).

Where did/do you point? Where is "ME"? It moves up a bit to your heart region, above Billy's stomach — simply a corporeal centralization within adult anatomy reinforced by psychological awareness of the indispensibility of the heart, reinforced by the emotions that live in language and art that refer always to the heart as being one's essence, etc.

Pretty fundamental though heady stuff. Nobody is pointing to his elbow or his leg!

If you were to ask this question of the Universal super-heavyweight boxing champion, he might reply, clinching and brandishing his right fist, "Right here, man." A full-time intellectual might not even articulate a reply but simply point to his brain or blink his eyes once with tolerance of the question.

In either such case, something would be missing: the human dimension.

And so it's fair to ask, "What is it that *is* where I am, who I am?" The questions that follow on rapidly are, of course, "Why *am* I", "for how long", "who's responsible?", "should this be pleasing to me or not?" etc.

Nothing of this is new, but all of it is eternal. Religion was formed to answer such questions people couldn't answer for themselves, to tie things together for significance and security. We presume that, within religious study (not just Christianity, mind you) the concept of the *soul* and the pactice soul-searching were born.

Even the most sophisticated etymological dictionaries cannot trace the word "soul" to a definite origin. It is obscure, just as the spark of life is obscure. Formal core-teachings equate the soul with the essence of being, the total existential awareness of being.

Although the soul is not a gland, a muscle, a complicated spiral of hormones and chromosomes, or a problematic growth, some so-called spiritualist practitioners in every age have tried to give it specific form. Interestingly, they would probably accept the proof of Billy's gesture or yours as validating the physical domicile for the soul. "Why it's the thymus gland, of course!"

And then Alan Leo, for example, that gifted, strange, yet catalytic astrologer of turn-of-the-century England, whose work influenced mightily the very beginning of astrology in America, keys us to another dimension. He made quite a statement about the soul in relation to Neptune (or perhaps I should say he made a statement about Neptune that involved the soul in a very insightful way). He said, "Neptune allows the soul to leave the body." Where is it going? Why? Where was it? Are we talking the sense of "journey" here?

What Leo meant specifically was that Neptune underlined the difference between the tangible and intangible realities of our existence. For Leo, the soul was intangible. It was an essence. In fact, when you get deeply into all this, you sense that this Leonian tie-up between Neptune and the soul was simply explaining the ultimate obscure dimension with the ultimate astrological dimension of that time.

For example, part of the heavy baggage Saturn has carried with it out of past ages into modern times was taken on as the last outpost of what astrologers could see and measure. Everything "left over" went to Saturn: all the residual pains, hidden fears, and environmental threats (thus all the study of different "levels", octaves, rulerships, etc.).

The discovery of Uranus freed things up, and the individual gained some respectability. But as well, the blame for everything left over from the symbolisms of those days was placed upon individualism disrupting society.

Then came Neptune.

Where is Jeff Green? is the question I asked myself as I watched this astrologer emerge in the world of astrology. I read some of his articles and interviews in several magazines, I attended some of his lectures, I asked around. And then, I discussed his new manuscript with him.

Jeff Green was pointing to Pluto, to everyone's Pluto, to everyone's "midsection" and saying it was where astrologers could begin to manage the mystery of soul, appreciate the sense of journey within evolution.

As I studied Jeff's manuscript, I enjoyed great relief: here was no spiritualist weaving spells. Rather, here was an extremely sensitive, spiritually eclectic, and psychologically sophisticated astrologer working with the intangible world in a tangible way.

Where attempts to discuss the soul in the past have always seemed to leave individualism behind (Leo's struggle between Neptune and Uranus), i.e., there's no way to follow the dictates of the soul unless you put those above the ways of the world, Jeff sensitively dichotomizes the "desires" structure within the soul, within "immutable consciousness": one desire is to separate from earliest origin (individuate); and the other is to return to the Source (be part of the whole).

Jeff then takes all of this insight and brings it down to earth through Pluto's occurrence in the Houses of the horoscope. Each placement is deeply, deeply studied in many dimensions, and each one is another finger-pointing for the concept of Soul: "Here *I* am."

A real test — As I was reviewing this manuscript, I received a particularly dramatic telephone call asking for astrological consultation. "George" was calling from a long distance, and his dilemma was extraordinary.

George seemed extraordinarily intelligent, lucid, exquisitely aware, informed, and disarmingly open-minded. Part of his concerns was that he knew emphatically through substantiated visions, dreams, and auditory encounters that, in a previous life, he had been one of the apostles of Jesus.

This was no laughing matter. Not even a smile could emerge in the telling of this state of affairs, so keenly circumspect was George in the discussion of his case. Indeed, such persuasiveness is often part of such a complex self-image, but to judge is not the role of the astrologer. Neither of us would have gained anything if I had given him the same doubting or flippant response he got whenever he dared to tell a friend about his problems.

His past-life identity gave him problems in present-life activity because of guilt and shame: he could not proceed with this life the way he was living it because everything he did appeared to be excruciatingly divergent from the standards established by the affiliation he had enjoyed in a previous life.

There were many other details, but this Preface is not the place for a case study. Upon accepting responsibility to try to help George help himself, I searched for a starting point. I needed to know more about the course of evolution within his life perspective.

Jeff Green's manuscript at that moment in time was 10 inches from my telephone. I restudied a certain chapter, and the discussion of

George's Pluto placement put me in touch with thoughts that turned out to be very helpful to George.

Jeff's work pointed out George to George. It does the same for *all* of us as we catch up with Pluto's accelerated journey through space and significance within time.

Congratulations, Jeff; and thank you.

Noel Jan Tyl
McLean, Virginia
September 1985

FOREWORD

BY ALAN OKEN

As I read and reread the pages of Jeff Green's fine, first book *Pluto: The Evolutionary Journey of the Soul*, I couldn't help but feel that an alternative title could be rendered to this edifying astrological work. That title is: "Astrologer, Come Home." Come home to the true reason for the practice of our ancient science and our incredibly beautiful art. Come home to the indwelling dedication to the service of our fellow human beings. Come home to the simplicity of the Life which underlies the Plan for the evolution of Humankind.

Astrology is an aspect of the metaphysical tradition known as the Ancient Wisdom. When correctly practiced, it leads the astrologer into a direct awareness of the demonstration of the "Geometry of God". It then reveals that there is a logical, provable patten to the unfoldment of human consciousness. Through his or her skills, techniques, and most importantly, his or her developed intuition, the astrologer becomes aware that there is an order to the seeming disorder in the world. What is more, that this order is measurable and predictable. Through the dedication to these studies, the astrologer then has the potential to become subjectively attuned to the understanding that this is a living consciousness. That indeed the universe is ALIVE and that this life has a Will, a purpose, and an intelligence, all of which are expressed in our solar system through the primary quality of Love. This Love is the true essence of all being and the ultimate "evolutionary cause of incarnation".

With the notable exception of that most authorative work, *Esoteric Astrology* by the Tibetan Master, D.K. (through His amanuensis, Alice A. Bailey) and the contributions of Alan Leo, W.P. Rigg, Mae R. Ludlum, the much-loved Isabel Hickey, and a very few others, there is a paucity of work on the more esoteric and spiritual facets of astrology. Dane Rudhyar has made one of the most significant contributions to astrology, not only in this century but in the whole of the recorded history of our science. He has outlined a philosophical frame of reference to bring to light the concept of the archetype of the Aquarian Age man and woman. This in addition to his developments and discoveries in terms of delineation of the natal chart! But even Rudhyar's vast work (and the highly significant books by Marc Edmund Jones) has not satisfied the needs and interests of the metaphysically-oriented astrologer, the seeker whose Love of God and humankind burns more brightly than the greatest of ideas and the most refined of intellects.

Jeff Green's current work (as well as his teaching and lecturing) touches the core of the metaphysical seeker and the spiritually-oriented, humanistic astrologer. Jeff Green's work is part of a larger, worldwide movement amongst astrologers to bring astrology out of a restrictive, glamour-riddled, personal power-seeking past. *Pluto: The Evolutionary Journey of the Soul*, helps this collective effort by focusing on Pluto, the planetary agent of transformation, in a way which, though esoteric in its orientation, is nonetheless accessible to all serious students of astrology.

Jeff Green's book has a unique way of combining metaphysical principles and laws, aids to the delineation of the natal chart, and distinct, clear insights into the workings of Pluto. This information is then presented to the reader in a language which is both reflective of Jeff's own metaphysical training and yet colloquial. His idiom, like Jeff himself, is conversational, friendly, yet deeply schooled in the classrooms of the Ancient Wisdom Teachings. Jeff is also a grown up "flower-child" and the profoundly committed humanism of the 1960's is sincerely reflected through his teachings and in this current volume. In these respects and in the arduous task of working for the reorientation of the personal life to the life of the Soul (and thus the well-being of the collective of Humanity), I am Jeff's brother. It is a joy to companion him and so many other brothers and sisters who share this orientation, on the Path of Service through astrology and metaphysics.

As humanity progresses further and further into the Age of Aquarius, the need to integrate personal Apirations through a collective focus of planetary participation will become increasingly apparent. It is by so doing that the new order of human expression will emerge. If enough Love underlies this process, such an emergence will not have to be made over the ashes of the destruction of our present world. Yet destruction is a necessary part of resurrection; death the needed vehicle for the consistent reincarnation which constitutes so much of the inner structure of our planetary and human life. The non-regenerative emotional and mental habit patterns, the physical and non-physical toxins and blocks, and all other sins of separation that are so much a part of karma, will have to be released so that the process of purification and rebirth may occur. As Pluto, the planetary agent of death (and rebirth), passes through its own sign of Scorpio, we will see an intensification of this urge for the destruction of the negative so that a greater future may manifest. As many of us know, this passage will take twelve years to complete and cover the years 1983-1995. Saturn, Lord of Karma, is also transiting Scorpio through the end of 1985 so that Jeff Green's fine treatise on Pluto couldn't be more timely.

We stand at the cusp of two ages. Scorpio is the "cusp" of two frames of reference for human expression: the personal and the collective. It is through the sign Scorpio and the activities of the eighth house of the natal chart that a tremendous test of transformation takes place for each of us. It is here that the individual dies to the urge for the attainment of personal desires (as initiated through Aries and the first house) and is forced either through spiritual aspiration or material and emotional desperation to give up the attachment to the personal ego. It is here, in fact, that the personal ego "dies" to the incoming impressions of the collective focus of the Soul. Libra, the previous test, has taught the meaning and the process of unification. Scorpio takes this one step further by allowing the now "other realized" individual to submerge into the waters of the collective life experience for the first time. The individual may then reemerge with the potential for the inner awareness of the mystical union of "self-other-All". This process does not culminate (nor does it have the greatest potential to contribute to the material world) until the individualizing Self is first bathed in the Fire of Universal Truth (Sagittarius). It is only then that such a developing Self may manifest through the initiatic force of Capricorn. It is thus very important to emphasize that Scorpio and the eighth house are but the processes of transformation leading up to this higher state. Since the Agent of Transformation (Pluto) is indeed transiting this "cusp of human consciousness" in Scorpio, the entire body of Humanity is currently undergoing the test of death and rebirth and will do so in increasing levels of intensity until the end of the millenium.

It is very interesting to note that the crises of human existence on the material plane (which always lead to questions of deeper moral significance) take place through the lower octave of Pluto and the co-ruler of Scorpio, Mars. Mars is traditionally known as the "god of war". It is very painful to observe that humanity is not more metaphysically trained and oriented. It would then be able to stop the wars before they materialize by being able to create transformation on the more subtle planes of consciousness before their horrific, material manifestations. It is usually through the threat of tribal, racial, national, and now, in our nuclear age, planetary annihilation, that the catalyst for the evolution or "devil-ution" of humanity takes place. It is also an historical fact that all too often the excuse for war is based on the need and desire for territorial acquisition and other economic circumstances. And where in the horoscope do these circumstances lie? They are found in and through the polarities to Scorpio, Mars/Pluto, and the eighth house: Taurus, Venus, and the second! There is an underlying cosmic order, demonstrable through the patterns of astrology. The chaos in the world is the product of the lack of awareness of the existence of such an order and the unwillingness of humanity to lovingly obey and further that order all the way to the Promised Land (called "Earth"). We continue to be such fools, such hungry children. We can and must change this unregenerative state of being. Consciously working with the energies of the planet Pluto, no matter how threatening to our egos, is certainly a way to manifest such essential growth.

Time and again we precipitate our nations into the suicides of war as the only phenomenon potentially big enough to knock us into some semblance of sensibility. But even these all too numerous outer wars make very little impression on the positive development of the collective consciousness. No, it really is and must be the inner wars of personal transformation which lead us to that place of peace. It is at this inner and sacred place that the individual ego is transformed by its meeting with the Soul. And it is the Soul that is the place wherein dwells the Master within each of us.

This peace is the one which "passeth all understanding" for such a peace does not issue from the mind nor from those spiritual leaders and heads of state who are exclusively mind-oriented. As Jeff Green says in this book: " . . . . . go slowly and let yourself meditate and feel the ideas". The "feeling of ideas" comes from the sensitivity learned through the identification of the ego with the Soul. It is only through such a process of essential unification that Humanity can be saved from its own total destruction. This unifying process is called "The Path of Discipleship" by the Ancient wisdom Teachings.

It is my deepest prayer, my sincerest hope, and the goal of the work I share with Jeff Green (and so many other brothers and sisters on this Path) that we declare total war on our lower selves (using the weapon of unconditional love). May Pluto's atomic bombs of love burst open the molecular structure of the little egos so that the true light of the Higher Self may manifest through us and seed the present and coming Humanity.

Alan Oken
Santa Fe, New Mexico
January, 1985

INTRODUCTION

This book is for the practicing professional astrologer whose client asks "Why am I here, and what are my lessons?" It is for the student of astrology and all of us who ask this question and try and answer it within the framework of astrology.

If we agree that astrology is a symbolic language or system describing the totality of life, then we will understand that this description is based upon empirical observation and correlations — correlations to the planetary pattern at the time of the observation. Three observable phenomena are evolution, cycles and growth. All are based upon process. Process involves a past, present and future. All of Creation, from a collective as well as an individual point of view, is governed by the natural law of evolution. There is an observable past, present and future as viewed from the past.

There are two kinds of observable evolution — cataclysmic and continual. Cataclysmic evolution occurs when a multitude of forces culminate at the same time to produce a total metamorphosis of that which was. The result will appear to be sudden change, but this is not the case. The result is based upon many evolutionary forces converging together which meet some other resisting force. The resisting force simultaneously represents the present and the past. The evolutionary forces creating cataclysm represent the future as a consequence of the past. When the evolutionary force for change becomes stronger than the resisting force, the result is cataclysmic change. The relative strength of both forces determines the intensity or magnitude of the cataclysm itself.

A simple example of this process at work in nature is an earthquake. In human society, the Watergate episode that led to the resignation of President Nixon is also an example of the same process.

By contrast, evolution through continual uniformity brings progressive or smooth cycles of growth that occur over great lengths of time. This type of evolution is much gentler and is not cataclysmic. It implies little or no resistance to the natural evolutionary cycles of life, individually or collectively, from a natural or human science point of view.

Pluto is the planetary symbol that correlates to the evolutionary process from a collective and individual point of view. It is the prime mover, first cause, or bottom line to which all other planetary factors are linked.

When the client asks "Why am I here, and what are my lessons?" the question implies the phenomena of evolution in the person's life. The ideas, methods, and techniques presented in this book offer suggestions as to how we can assist the client with these types of questions. This approach is simply one more filter to put on our astrological eyeglasses to help us see the answer to these questions of personal and collective evolution.

This approach will help identify the past evolutionary context of any individual, why he or she had that past, and how that past has conditioned the individual to the present state of conditions in his or her life. By understanding the evolutionary intent of the Soul, we will be able to identify the evolutionary lessons that are represented in this life for any individual. The current life evolutionary intent or lessons allow the individual to grow and evolve into the process of his or her own becoming; to grow and evolve in some way at every moment in time. By understanding the past evolutionary context, we will understand the reasons or the "why" of the particular lessons suggested in any birthchart. By understanding the collective evolutionary past (humanity) we will be able to understand the collective present, and the choices that humanity faces that will create the future. More than this, we will understand how the past created, shaped, or conditioned the specific choices that we face in every present moment, and why we have those particular choices.

This kind of individual and collective understanding is extremely important now that Pluto is in Scorpio, and Neptune is in Capricorn. This particular cycle only occurs every five hundred years or so. It always occurs just before, or just after, the turn of a century.

If we turn to history in order to examine the evolutionary and karmic themes that have happened when this cycle has been in force, we find that the essential choices and issues revolved around unity or polarization. If we meditate on these two archetypal themes and project them into today's world with respect to the issues at hand, the consequences of the individual and collective choices that reflect these themes will be seen to be more critical than ever before in human history. The number one issue in our world today must be the collective approach and response to nuclear technology, and the military weapons that it has spawned. Pluto correlates to the core, nucleus and penetration and fusion of anything — the basis of the nuclear and related bombs. It also correlates to the phenomena of defense, or defensiveness, for whatever reason. Pluto also corresponds to how we use the resources of ourselves and others.

With Pluto now in Scorpio, these natural correlations, and the issues that they represent, are clearly in the process of being magnified and intensified. The polarity point of Scorpio is Taurus. Taurus represents in its deepest and most essential archetype (from a collective and individual point of view) the instinct for self-preservation; the instinct to survive. How will the choices made with respect to Pluto in Scorpio impact on the survival of the human species?

History points out that nationalism, authoritarianism, conservativism, purposeful deceit issued by those in power in order to maintain that power, religious events in one form or another (Jesus was on the planet and crucified with Neptune in Capricorn), individual and collective disillusionment and/or revelation, and the dissolving of preexisting and outmoded structures of reality, has occurred when Neptune has been in Capricorn. This dissolving effect creates a sense of uncertainty in the collective Soul which promotes the need to believe in the illusion that everything is fine. This collective need translates into an attraction to those who tell them what they want to hear even though it is not based on reality. Commonly, scapegoats in one form or another are found to blame the ills of the day upon. They are persecuted, attacked, or in some way isolated from the "mainstream" who want to conform and believe that everything is fine. This phenomena occurs internally within a nation, and externally as one nation, or a block of nations, isolate another or other nations. These facts of history are amazing in light of the archetypal intent of Neptune in Capricorn: to transcend the boundaries of national identity and strict self-interest, and to individually and collectively unite with the Spirit, Force, or Source. This Source created all things so that mutual cooperation can occur at all levels of reality to solve the problems and issues at hand so that the survival of the individual and the human species can continue.

The story that is told throughout recorded history when these five hundred year cycles have occurred is not pleasant, nor particularly inspiring with respect to what must be done in our world of today. Generally, polarization of peoples and nations has occurred rather than the necessary unity. Segments of the population have drawn lines in the dirt that staked out their particular points of view, self-interests, value associations and belief systems. These groups then challenged, or attempted to impose their views on any other segment that did not conform. Commonly, religion or some righteous moral imperative has been used to justify or defend these actions. In effect, one segment thinks they are right and that all others are wrong. Nations have experienced intense polarization and conflict with other nations for the same reasons, under the delusive name of national interests. Commonly the leaders of the different governments play upon the theme of nationalism during these cycles.

The archetypal need to regenerate through transcendant unity has rarely occurred. Instead degeneration through polarization and power confrontations has dominated the times when Pluto has been in Scorpio, and Neptune in Capricorn. Interestingly enough, Pluto is always within the orbit of Neptune during this cycle. The two hundred and forty eight year cycle that measures Pluto's return to Scorpio correlates to the culmination of the evolutionary cycle that began at its last return, and the beginning of a brand new evolutionary cycle reflected in its current passage through Scorpio. That which is culminating is based on all the prior actions, negative and positive, that have occurred all over the planet during this two hundred and forty eight year cycle. In other words, the results of all those prior actions are coming to a head. This culmination implies that a new evolutionary cycle is beginning so that the survival of the planet and human species can continue. Pluto and Scorpio both intensify and concentrate whatever dynamics that they are linked too. Thus, Pluto's passage through Scorpio "intensifies the times." This intensity is necessary for it enforces awareness of the individual and collective issues that must be confronted, eliminated, or redefined in order for individual and collective growth, evolution and survival to continue.

The tension produced during the transition between the old and the new evolutionary cycles can be quite intense. One segment of the population will attempt to maintain the old order of structural reality, another segment will attempt to move forward with respect to defining the issues that must be confronted and solved in new ways so that a positive and necessary growth can occur, and another segment will be more or less ambivalent toward the concerns, issues and claims of these two polarized segments. In effect, one third of the poulation is lined up along the lines of the past, one third along the lines of future necessities, and one third is ambivalent. This individual and collective state of affairs only promotes and reinforces polarization, not the necessary unity that will allow for a smooth transition with respect to the evolutionary necessities of the times.

Today the issues and problems are many. These issues include nuclear war, the delusion of nationalism and perceived national self-interests, the interdependant national and international banking/economic structure, the use and distribution of national resources in all governments, food and food production, how much and where the governments commit their money, trade issues with respect to imports and exports, segments of the population becoming dislocated through job loss due to technological advances (the last industrial revolution began with Neptune and Uranus both in Capricorn, and intensified when both these planets entered Aquarius and Pluto moved into Aries), the use and application of genetic engineering, and, above all else, how to promote a global consciousness in the citizens of the world.

   This last point is interesting to consider because in reality most people do have this level of awareness in varying degrees due to the media. The common person in all lands truly desires peace and an extension of goodwill toward all. Yet, the people who are in power in the dominant governments of this world are creating policies that lead in just the reverse directions, while paying lip service to the concerns of the people that they lead. This apparent dichotomy is exactly where the lessons of history should be learned. It is this dichotomy that has led to the major problems of the past when Pluto has been in Scorpio, and Neptune in Capricorn.

   If we are to learn from history so that we do not repeat the past mistakes, the people in all lands must make their voices and desires heard. When necessary and possible, the people must transcend governmental policies that promote polarization through direct action. A contemporary example of where the people have overridden the policies of a government to promote a transcendental unity is the current tragedy of Ethiopia. The famine that is now occurring (as of this writing) was triggered when transiting Pluto moved through Ethopia's eighth house in direct opposition to its natal second house Venus in Taurus. Obviously the major governments of this world did not react quickly enough even though many knew of this probability well in advance of its actuality. Even after the news leaked out to the world, governments like the U.S. did not respond because Ethiopia was a Marxist government. Once the dimensions of this tragedy became public some governments did respond, including the U.S., but not in a way that was adequate to solve the problem. Yet, the voluntary efforts of many musicians in America, Great Britain and South America, along with many other private donation efforts, have helped toward relieving the problem as well as toward promoting a transcendant and universal social consciousness that goes far beyond governmental policy. Even the title of the album that the musicians in the U.S. produced suggests this necessary evolutionary theme: *We Are The World*. If you listen closely to the lyrics in this song you will hear the evolutionary themes: "We are the world, we are the children, we are all God's great big family, we all must lend a helping hand, the choices that we are making will save our own lives, we are the one's who make a better day, just you and me." This message has been responded to by millions of people in many lands. Let's hope this example of the common citizens of the world impacting in a positive way will continue to advance the evolutionary requirements of our times.

   The individual and collective choices that we all make over the next fifteen years will perhaps be the most critical choices that humanity has ever faced, determining the fate of human history. The weapons of destruction that Man now possesses makes this so. If we are to negate the possibility of some of the darker scenarios that could occur, the average citizens of this world, you and me, will have to demand that our governments move toward the spirit of unity, reconciliation and cooperation with all, rather than toward a policy of national and international confrontation and polarization. We are the world, we are all part of God's big family. In this spirit, the choices that are made can only lead to a positive, healthy and productive evolution for this planet.

It is important to remember that the evolutionary requirements of our times *will* occur in one way or another. They will occur either through cataclysmic evolutionary and karmic events, or they will occur through slow, yet progressive events that reflect the proper choices that must be made by all of us: as individuals and as a collective body. By confronting and understanding the issues, the proper choices preclude the necessity of cataclysmic change. It was in this spirit, by the way, that Nostradamus offered his visions of the future. He felt, as does the author, that the cataclysmic events as predicted by so many for the immediate future are not fated. Nostradamus was born just after the last cycle of Pluto's and Neptune's passage through Scorpio and Capricorn respectively. In fact, Nostradamus had Neptune in Capricorn, and Pluto was a few degrees into Sagittarius in the ninth house. If the major portion of humanity, and particularly the countries that hold world power, continues its existing course toward nationalism and polarization, the cataclysmic events predicted by so many has a high degree of probability. The necessary changes will take place.

Toward the end of this century Neptune and Uranus will move into Aquarius, and Pluto will move into Sagittarius (a fire sign). A simple review of history tells the story of cultural and societal restructuring when this pattern happened before (the Industrial Revolution, among other events, of the 1820's and 30's). This restructuring pattern has been and will be accelerated during this cycle. The choices that are made while Pluto is in Scorpio, and Neptune in Capricorn (to be joined by Uranus at the end of this decade) will determine how this societal and cultural restructuring is experienced by all of us.

The emphasis in this book is on the individual because that which occurs collectively starts with the individual. If each person understands what his or her own evolutionary requirements are, and operates in such a way as to actualize those requirements, then the collective evolutionary necessities will be developed in a noncataclysmic way. This book was written in this spirit.

Pluto will be linked to the concept of the Soul, and how the phenomenon of individual evolution originates from the Soul — the prime mover, or "bottom line". We will be discussing what the Soul is, and the inner structure or dynamics of the Soul that serve to propel its evolution. We will also be discussing the issue of free will or choice; not only how this phenomenon works, but how it is the basis of the two types of evolution. The evolutionary past of any individual obviously implies the phenomenon of reincarnation: the transmigration of the Soul.

The evolutionary past of any individual also represents the deepest sources of unconscious emotional security. These old unconscious security patterns create resistance to change and the unknown in varying degrees of intensity in all of us. This security/resistance dynamic creates compulsion, obsessions, or emotional complexes in many individuals (or nations). This book will explain how these compulsions and complexes are rooted in and linked to the past. We will discuss the idea or phenomenon of karma: what it is, what it is not, and how it is linked to the evolutionary factor for each of us. Numerous chart examples will show how the evolutionary factor works and will illustrate the principles in this book. This work will be divided into two volumes. Volume One will cover the essential ideas, methods and techniques as well as Pluto through the houses, Pluto in aspect to other planetary factors, and Pluto in transit, progressions and solar returns. Volume Two will be based on Pluto's evolutionary and karmic role in relationships that will involve both the composite chart and cross chart placements (synastry), Pluto's correlations to anatomy, physiology and the chakra system (a complete discussion of what the chakra system is and how it works will be presented), and Pluto's evolutionary role through modern history will be gone into in depth with some possible projections relative to the future based on that which has occurred in the past.

Because the focus of this book will be on the evolutionary intent and necessity to grow and change, the description of Pluto through the houses and signs will involve old patterns, orientations, and behaviorial manifestations that must be metamorphosed so that growth and evolution can occur. This focus implies preexisting limitations and imperfections. Some readers may feel that these descriptions are "negative". However, they are not intended to be negative. The essence and nature of evolution is change. It implies limitations or structures that are blocking necessary change. The descriptions of Pluto through the houses and signs will objectively describe what common limitations and imperfections must be encountered for the ongoing evolutionary intent to occur so that perfection can be realized at some point in the evolutionary journey of an individual. This book is not intended to be an astrological "cookbook." This book should be read all the way through so that a total and comprehensive understanding can be developed as to the nature and basis of the principles expressed herein.

I hope these ideas will help lead to an astrological understanding of "Why am I here, and what are my lessons?" — both individually and collectively. Because of the nature of these questions and the nature of Pluto, the reading may be intense at times. I would suggest that you take your time with this book; go slowly and let yourself meditate and feel the ideas. The ideas and principles are not new. They are as ancient as civilization itself but are now linked to astrology in a new way. I sincerely hope that you will enjoy this book.

God Bless,
Jeff Green
Seattle, Washington

CHAPTER ONE

PLUTO: THE EVOLUTIONARY JOURNEY OF THE SOUL

Pluto correlates to the Soul and evolution. What is a Soul? According to many spiritual, religious and metaphysical sources, including the Bahagavad-Gita and the Bible, the Soul is an immutable consciousness that has its own individuality or identity that remains intact from life to life. In each lifetime, of course, the Soul manifests a personality that has a subjective consciousness and unconsciousness. Saturn defines the boundaries of our subjective consciousness — that of which we are consciously aware. Uranus represents the individualized unconscious, Neptune the collective unconscious and Pluto the Soul itself. Each personality that the Soul manifests, from lifetime to lifetime, has an ego (Moon) that serves as a focusing agent to create one's own self-image. The Moon is similar to the lens in a movie projector — without the lens there would be no picture upon the screen. Each personality has an intrinsic nature, an orientation that experiences life in its own unique way. Each personality that the Soul manifests relates directly to the evolutionary necessities of the Soul. The Soul through the personality must experience life in particular ways in order to grow and evolve. Each personality created is directly linked to the past evolutionary and karmic history of the Soul.

Within the Soul there exist two coexisting desires. One desire is for separate existence — to separate from that which created the Soul. The other desire is to return to the Source of creation. The interaction of these two apparently opposing desires instigates the drama of personal and collective evolution. Desire is the determining force that dictates the reality of each individual. Buddha's enlightenment underneath the bodhi tree was based on just such a realization as he pondered the nature of sorrow, pain and misery. Planetary symbolism is interesting to consider here. Pluto is a binary planetary system. Pluto's moon, Charon, is actually a planet in its own right. Charon is half the size of Pluto itself, and one twentieth the distance from Pluto as compared to the distance of the Moon to

1

Earth. The principle of dual coexisting desires seems to be reflected in this planetary symbolism.

The interaction of these desires determines what we think we need. That which we think we need determines the choices that we make. The choices that we make determine the actions that we initiate. The actions that we initiate determine the reality that we experience. Actions lead to reactions which lead to other actions. Simply stated, this process is the basis of what is called karma. We reap what we sow.

It is because of these dual archetypal desires that the story of personal and collective evolution, through countless lifetimes, is so long. It would seem clear that the Soul's evolution is principally based upon the progressive elimination of all separating desires. As these separating desires are eliminated, the desire for return to the Source becomes stronger and more dominant in our consciousness. The operation of these dual desires is an easily verifiable and observable phenonmenon. Is it not true that in each of our lives we manifest all kinds of separating desires, i.e., the new possession, the new lover, etc., etc., etc.? And, is it not also true that after we experience the momentary pleasure or high" of obtaining what we desire, that this satisfaction is soon replaced with the old familiar sense that there must be more than this — the object of the desire? This sense of "something more" only fuels another separating desire. As we evolve over countless centuries we begin to exhaust or eliminate these separating desires. As we do so, the desire to know that which created us becomes stronger and stronger. One day it is the only desire left in us. Ultimately, this evolutionary process produces Creation-realized Souls such as Jesus, Buddha, Lao-Tzu, Yogananda, Moses, and other great saints and masters.

> "Mankind is engaged in an eternal quest for that 'something else' he hopes will bring him happiness, complete and unending. For those individual Souls who have sought and found God, the search is over, He is that something else."
>
> Paramahansa Yogananda
> *Man's Eternal Quest*

It should be clear that the desire to return to the Source is the stronger of the two archetypal desires. Yet the fact that there are two coexisting desires accounts for the apparent phenomenon of free will, or free choice, and the phenomenon of decision making. This story is found in all religious doctrines, including the Bible, as the story of Adam and Eve. So, yes, we are responsible for our own evolution, for our karma and the experiences of reality that we all share.

"Why am I here?" and "What lessons am I learning?" are questions that we all ask at some point in our evolutionary cycle. How are we to answer such questions using the birth chart? How are we to answer them from a point of view that is applicable to all individuals? Many people could care less about karma, the Creator, or understand-ing themselves in a holistic or universal context, yet they still want to know what they should be doing. As Carl Jung pointed out, the prime role of the counselor is to validate objectively the subjective reality of the client. As counselors, or just as friends, we will do our best if we can simply identify reality as it exists for those with whom we interact, and to give them what they need according to their own reality — not ours. This type of giving requires observation, listening and objectivity. The very language that we use is based on their reality, their language. So as we attempt to answer the questions "Why am I here?" and "What are my lessons?" we must attempt to understand the "reality" implied in the birth chart.

PLUTO'S HOUSE, SIGN, AND POLARITY POINT

Pluto's house and sign placement describe two simultaneous phenomena. On the one hand, the natal position of Pluto describes the generational vibration that a person comes in with, as well as the specific individualized patterns in identity association implied from the evolutionary past: the desires, beliefs, thoughts, perceptions, values and orientation to reality itself. On the other hand, the natal position of Pluto points to the evolutionary desire, intent, or cause of this life as seen in Pluto's opposite house and sign.

Whether the individual is conscious of this intent or not does not matter because the Soul, not the personality, is the ultimate causal factor, or determining force, behind each life. The lesson will occur in some way through evolutionary necessity. If we are conscious of the intent for this life we can foster it through cooperation and non-resistance. As the life unfolds the inherent patterns of identity association, seen in the natal position of Pluto, will be reborn to a new level of expression described by Pluto's opposite house and sign. The implied limitations of the past will be experienced in some way so that the evolutionary point will transmute these limitations into new levels or horizons of awareness and expression.

When we link these principles to the coexisting desires inherent in the Soul, we will see that there is an implied tension between the past, present and future. The potential for internal conflict is enor-mous. The desire for separateness can lead to an effort to maintain the past, while the desire for return to the Source promotes a focus-ing on the future (the polarity point of Pluto). The tension or conflict, if there is subconscious resistance, occurs in each moment of our lives.

"When you arrive, or emerge, into physical life, not only

is your mind not a blank slate, waiting for the scrolls that experience will write upon it, but you are already equipped with a memory bank far surpassing that of any computer. You face your first day upon the planet with skills and abilities already built in, though they may not be used; and they are not merely the result of heredity as you think it . . . you may think of the Soul or entity . . . . as some conscious and living, divinely inspired computer who programs its own existences and lifetimes. But this 'computer' is so highly endowed with creativity that each of the various personalities it programs . . . in turn creates realities that may have been undreamed of by the computer itself.

"Each such personality . . . is highly tailored to meet very specialized environments. It has full freedom but it must operate within the context of existence which it has programmed. Within the personality, however, is the condensed knowledge that resides in the computer as a whole.

"Each personality has within it the ability to not only gain a type of existence in the environment — in your case in the the physical reality — but to add creatively to the very quality of its own consciousness and in so doing to work its way through the specialized system, breaking the barriers of reality as it knows it. . . . . "

Notes from Seth Speaks
Metaphysics of Counseling
Autumn, 1977. Page 40

Some simple examples will illustrate these principles. The sign that Pluto is in, i.e., Pluto in Leo, represents the generational vibration that we come in with. Pluto was in Leo from 1937 to 1958. Obviously, millions were born with Pluto in Leo. In general, this generation has natural desires and pre-existing orientations for creative self-actuali- zation. Pluto in Leo individuals need enough freedom to actualize, in a creative way, their own unique purpose for living. This need implies a deep self-orientation, a more or less narcissistic orientation to life. Pluto in Leo implies a creative generation that needs to be in charge of their own lives, and able to take destiny by the hands and create it out of the strength of their own wills. Yet, the polarity sign for the generation is Aquarius. As a generation then, they must realize Aquarian lessons that involve developing an awareness of the whole, of humanity, versus an awareness of just themselves. By developing this awareness, the generation could then creatively realize their specific purposes by linking their work or identities to the socially relevant needs of the whole. This process demands evolving an attitude of objectivity versus subjectivity, of detachment versus over-involvement with personal concerns. We are rapidly moving toward the Aquarian Age. Is it only coincidence that the Pluto in Leo genera-tion will progressively assume positions of social power in the next few decades?

In general, all Pluto in Leo people must learn Aquarian lessons. The specific house placements of Pluto will begin to indicate the specific and individualized lessons for each person. An example: Pluto in Leo in the Ninth House. Let's say this person, in a prior life, desired to understand life in a cosmological, metaphysical, or religious context by intuitively sensing that there was more to reality than meets the tactile senses. Commonly this individual, from a prior-life point of view, needed to initiate a wide variety of experiences in order to discover or realize the knowledge that she or he was seeking. This need or desire became part of the individual's own creative self-actualization process from an evolutionary point of view. The individual would come into this life with not only a highly developed intuition, but also a rigid organization of reality (beliefs) that reflected the desire for self-understanding in a cosmological sense. The polarity point would be the Third House and the sign Aquarius. Simply stated, the general Aquarian lessons mentioned above would be learned through Third House kinds of experiences. Thus, the individual would necessarily experience internal and external philosophical and intellectual confrontations in order to realize the limitations of his or her own organization of reality (beliefs). This does not necessarily imply that those beliefs are wrong, but that they are limited. The individual would progressively learn that the paths to truth (Ninth House) are relative, and thus learn to communicate in a variety of ways depending on the people or circumstances encountered. Keep in mind how the two coexisting desires in the Soul, and the potential tension or conflict can impact on this illustrated process. As an example, it would be very common for this individual, coming into this life, to defend the personal beliefs and the organization of reality relative to those beliefs. The individual commonly would attempt to convince or convert other people to his or her own point of view. The need to defend reflects the desire to maintain separateness. On the other hand, the desire to evolve, or grow, leads toward the intellec-tual or philosophical confrontation implied in the Third House polarity. The interaction of these dual desires, and the resulting behavior, will be a primary theme underlying the individual's entire life.

The potential for tension and conflict reflected in these dual desires of the Soul can be equated with the basic psychological phenomenon of attraction and repulsion. We can be attracted to the very thing that we are repulsed by, or we can be repulsed by the very thing that we are attracted to.

In the above example the Soul intends or desires to evolve. This is demonstrated, again, by the polarity point of the natal position of Pluto. As this desire reflects itself through the consciousness of the individual (ego), the individual is attracted to intellectual or philosophical discussions. Yet this very attraction will guarantee, because of the implied limitations from the evolutionary past, intellectual or philosophical confrontations with other people. The individual will experience his or her own philosophical or intellectual limitations because of these confrontations which will ignite the dynamic of repulsion through these experiences. In terms of behavior, the individual will alternately attempt to convince or convert others to their personal point of view, or withdraw altogether from this kind of experience to avoid dealing with this perceived and experienced conflict. Withdrawal creates an internal volcano of confrontation that induces intuitive thoughts or ideas that directly challenge the pre-existing beliefs or philosophical orientation that the individual brought into this life. Because the Soul (Pluto) desires to evolve (Third House) it will lead the individual into conversations with others in order to effect the necessary challenges that allow for a transmutation of existing beliefs. In other words, the individual will subconsciously be attracted to, and thus draw to them, those kinds of people who will have this effect upon his or her beliefs and philosophical orientation.

As this process unfolds throughout life, the individual may come to realize the evolutionary lessons involved and acknowledge the relativity of his or her own point of view. At this point they could participate in philosophical and intellectual discussions in a non-defensive way, learning from as well as teaching others. The individual would learn to understand and respect individual differences in beliefs, philosophies, religions and intellectual organization. Conversely, the individual may succeed in repulsing the evolutionary pressure or intent. In this case, the behavior would manifest in defensiveness and rejection of anyone who did not agree with his or her own point of view.

This attraction/repulsion dynamic linked with the dual desires inherent in the Soul will manifest in any natal house position of Pluto. Allow yourself to take a few moments to reflect on your own natal position of Pluto to see this principle operating in your own life. From an *ultimate* point of view, repulsion will automatically be a consequence of any desire and attraction to anything other than to return to the Source.

From a purely psychological point of view, Pluto correlates to the deepest emotional security patterns in all of us. These security patterns are unconscious. Most of us automatically gravitate to the path of least resistance. The patterns in identity association that are carried over from the evolutionary past are directly linked to the path of least resistance and, therefore, to our security needs at an unconscious level. The past represents familiarity and that which is known. Our lessons, or the evolutionary intent described by Pluto's polarity point, are not known. They are the unknown, the uncharted. The unknown as described by our evolutionary intentions directly challenges that which is known or familiar, and therefore challenges our security at the deepest possible level — the Soul, our core. How many of us are aware of our security needs at an unconscious level? How many of us are aware of how these patterns and needs control or dictate our behavior? I think most of us would agree that many of us are not. If this is true, then how many clients are aware of these patterns? Again, not many. These deep-seated security needs drive us to approach certain areas in life in exactly the same way over and over again. These areas are specifically linked to the natal position of Pluto. If we remain stuck in the old way (the past) for security reasons, problems will arise in these areas because the Soul in each of us desires to grow and evolve.

The natal position of Pluto represents an area of natural gravitation for security reasons and is thus given tremendous power. Pluto implies a compelling force to maintain that area (house and sign) in the familiar ways of operating — the past. This compelling force can be called compulsion or obsession if the resistance (desire) is strong enough. The strength of the resistance determines the magnitude of problems or confrontations experienced in Pluto's natal house and sign position. Because the evolutionary force of the Soul is to grow, the individual will either experience cataclysmic growth (evolution) because of the problems experienced through resistance to the evolutionary intent, or slow, yet progressive growth because of the relative non-resistance to the evolutionary intent.

These areas of natural gravitation will be the stumbling blocks to the individual's development and growth. The intense degree of power and identification of Pluto (the Soul) is the cause of the problem. Because Pluto represents an essentially unconscious process, the individual is basically unaware of the motivational patterns of the Soul, and may feel stagnated, frustrated, or wonder why the same lessons, the same mistakes, the same kind of relationships, the same problems occur over and over again. Eventually the person may ask the question "Why am I here?", "What am I supposed to be doing?", and "What can I do about these things?" These questions are a natural response to the desire for ultimate identification with the Creator, although the individual may not recognize it as such.

At this point, the astrologer can indicate ways to help the individual become freer and more growth-oriented. Once the person becomes aware of the problems, motivations, needs, desires, attitudes and security issues dictating his or her experiences, the necessary changes may occur — if the individual desires to initiate them. Everything begins with desire. Again, the path to change rests in the polarity point of Pluto by house and sign. As change begins to occur there will be an automatic redefinition, evolution, and reexpression in the natal house of Pluto. The individual will approach that area differently and will be reborn at a new level through death of the old behavioral patterns.

We can take these ideas about Pluto and apply them to anyone. Through observing and listening to the client, and analyzing the chart, we can determine the reality of that person as it exists for them. The ways in which we communicate or express these principles to another person are determined by the individual's level of understanding.

It is extremely important to understand a person's evolutionary past. In the many years that I have been doing astrological charts I have observed that seventy-five to eighty percent of any individual's behavior is directly conditioned by the past. Even modern psychologists refer to "unconscious forces" or "memories" that influence the thoughts, feelings, moods and desires in our subjective consciousness. It is important to understand the past because it explains the here and now, why this kind of life, and for what reasons. It explains everything from the point of view of evolutionary necessity and karmic causes. Why this parent, this lover, this experience, why these conditions in my life, and so forth. By understanding that we are in charge of that which we have been, are, and are in the process of becoming. We have choices at every step, and at every moment, in our lives. The choices are reflected in our desires, and desire determines everything that occurs, individually and collectively. By understanding life in this way, we can more thoroughly understand the "lessons" that we must learn. We can understand the "why" of those lessons, and by so doing, create a willingness to accept, not resist, these lessons because of the knowledge gained about the past causes. If you are a student of astrology these ideas can lead to the objective understanding of these issues in your own life. If you are an astrological counselor, then you may be able to use these ideas on your client's behalf.

Why is it a common experience to have a difficult time during the first Saturn return? The first twenty-eight to thirty years of life is spent living out the conditions represented by the evolutionary and karmic past. As the Saturn return approaches, the individual commonly begins to feel restricted, frustrated, or depressed by the reality or conditions of his or her life. The Saturn return is a natural growth period that is a condensed reflection upon the past and, therefore, accelerated because of the new evolutionary cycle that is in the process of emerging. As Saturn comes closer and closer to its natal position, the individual experiences two simultaneous states — a sense of restriction, frustration and depression based upon the old conditions (karmic heredity), but also the emerging instinctive sense of needing to redefine and recreate the life conditions. In other words, the past and future (the evolutionary intent of the Soul) begins to collide in each moment of the individual's life. As the evolutionary forces converge upon the existing reality (security) that is defined by the past (Pluto), the individual is presented with extremely important choices that will determine the experiences of the next twenty-eight years.

Of course there are other planetary cycles that influence this natural evolutionary process. Saturn has a natural seven-year cycle within the complete twenty-eight year cycle. Jupiter has a twelve-year cycle of growth and expansion. The nodal axis of the Moon has a natural eighteen-year cycle, Uranus forms a square to itself at twenty-one years of age, an opposition between thirty-nine and forty-two, another square at sixty-three, and returns at approximately eighty-four years. Neptune forms a square to itself at forty-two years of age. Depending on its ecliptical path, Pluto can also form a square to itself during the life of an individual.

During these natural cycles the evolutionary forces relative to the future are accelerated relative to the conditions represented by the past. Those evolutionary forces represented by the past define the total reality of the moment, both individually and collectively. Creation is in a continual process of birth, death and rebirth. We can observe this natural evolutionary process all around us — from the changing seasons to the solar system and universe beyond. So it should not be so surprising to consider that all of us are part of this process. To understand the past is to understand the moment. This understanding allows for positive choices relative to our individual and collective futures. We can help ourselves, and we can help humanity and the Earth in doing so. The natural cycles of evolutionary acceleration allow for maximum growth opportunities. If those opportunities are resisted because of the cumulative forces of the past, then the potential for cataclysmic phenomena exists.

PLUTO AND THE NODAL AXIS

Most astrologers have learned that the South Node of the Moon represents a composite of issues, i.e. value systems, beliefs, needs and so forth from the past that are relied upon or gravitated to in this life. We have also learned that the North Node of the Moon represents lessons to be learned in this life. A few professional astrologers insist that this correlation is wrong, that the North Node is the past and the South Node the future. Despite the insistence of these few, historical observations over centuries validate the correlation of the South Node with the past and the North Node with the future. Based on my own observations over many years, and reading thousands of charts, history seems to be correct. Meditate on these symbols in your own chart, and then reflect on your own life, and I think this correlation will be borne out.

The relationship between the North and South Nodes and Pluto is very important. We have correlated Pluto to desires based on evolutionary factors. The natal position of Pluto describes the pre-existing patterns in identity association that the individual naturally gravitates to in this life. The South Node describes, from a prior-life point of view, the mode of operation that allowed the individual to actualize or fulfill the desires or intentions described in the natal position of Pluto.

A simple example using the Ninth House Pluto discussed earlier will illustrate this point (see Sample Chart). Let's put the South Node in the Seventh House. The individual's desire to understand his or her life in a larger cosmological context would be the essence of the life. With the South Node in the Seventh House, how would we expect the person to actualize or fulfill that desire? From a prior-life point of view, the mode of operation would have led to the initiation of many kinds of relationships through which the individual was seeking the knowledge implied by the Ninth House Pluto. Relationships would typically be formed with others who could serve as "teachers" to the individual. These prior patterns imply a dependency on that kind of orientation coming into this life. Thus, the person would tend to draw, through relationships, others to them who would attempt to convince or convert that person to their points of view. In this way, the individual would have collected a variety of teachings from a variety of people. Which one is right, which one wrong? Is one more right than the other? Are they all wrong? These are questions this person might now be asking. In this case the North Node, the mode

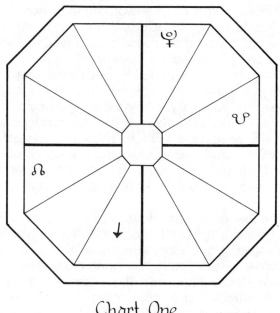

Chart One

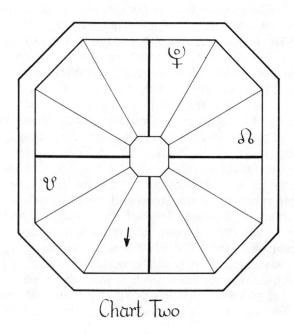

Chart Two

of operation, would be in the First House, and Pluto's polarity point in the Third House. What are the evolutionary lessons here? Simply stated, the individual would acknowledge the relativity of varying points of view, yet by using the mode of operation symbolized by the First House North Node, learn to develop his or her own voice, own identity, own vision, own knowledge, and would learn to ask and answer questions from within the self. This transformation would then progressively, yet totally, change the approach and orientation to relationships. Rather than depending on teacher-type relationships, they would subconsciously want to become the teacher. Further down the road the individual would desire a situation of equality in a relationship. Both partners would then be simultaneously student and teacher to each other by encouraging mutual independence in each.

Let's take the same Ninth House Pluto and put the South Node in the First House. The polarity point of Pluto remains in the Third House. The mode of operation to actualize or fulfill the evolutionary intent of Pluto is the Seventh House North Node. The bottom line is still the Ninth House Pluto and the desire for the individual to understand him/herself in a larger cosmological context. From a prior-life point of view, the individual fulfilled that desire by maintaining independence and freedom from involvement in relationships in order to initiate whatever experiences that he or she deemed necessary in order to fulfill the desire for cosmological knowledge. The individual was dependent only upon the self. They would come into this life with the ability to ask and answer his or her own questions. It is from exactly this orientation that the evolutionary limitations and intentions are reflected. Yes, the individual is learning the lessons of intellectual and philosophical beliefs symbolized by the Third House polarity point of the Ninth House Pluto. Yes, the individual is destined to experience intellectual or philosophical confrontations with others in order for this lesson to occur. How will this lesson occur specifically? Through the Seventh House North Node. The Soul of this individual would desire and need to be in relationships in this life. The individual will also be required to give to others, and must learn how to listen to other people in order to understand and identify another's reality as it exists for them. In this way they will be giving to another what is needed from the other's point of view (North Node in Seventh, polarity point of Pluto in the Third House) versus giving to others from the individual's own reality (South Node in the First, Ninth House Pluto). Commonly this individual, when disposed to be involved with people at all from a prior-life point of view, would have been the teacher. They would have needed to be perceived by others as a leader, as special, because of the knowledge that the individual presents (Ninth House Pluto, South Node in the First). Because Pluto's Third House polarity point will lead to philosophical or intellectual confrontation in order to expose the limitations of the individual's own belief structure, the Seventh House North Node creates a mode of operation through which the individual will also be taught by other key individuals on a cyclical basis throughout life.

Gradually, through confrontations, the individual will learn that he or she can learn from others. In addition, the individual will come to realize that he or she must learn how to be in relationships as an equal, that relationships are essential to his or her evolutionary growth. With this understanding, the individual will also realize, just as the person with the South Node in the Seventh House did, that equality will occur through encouraging the independent self-actualization of the partner.

The South and North Nodes thus correlate to the modes of operation that help the individual fulfill the evolutionary necessities and desires. The Ninth House Pluto was used in both examples. Yet the modes of operation, past and present, needed to fulfill those Ninth House desires were diametrically opposed. This principle can be applied to all combinations of Pluto and the nodal axis.

The next principle that we need to consider is this: What signs and houses do the planetary rulers of the South and North Node occupy? The planetary rulers of the nodes are the planets that "rule" the sign that each node is in, i.e. with the South Node in Libra the ruler would be Venus. The house and sign position of each planetary ruler of the nodes act as facilitators that the individual has either used to develop the mode of operation from in the past, or to develop the mode of operation relative to the evolutionary intent in this life.

An example: The Ninth House Pluto in Leo, South Node in Gemini in the Seventh House. The ruler of Gemini, Mercury, is in Pisces in the Fourth House. The North Node is in Sagittarius in the First House. The ruler of Sagittarius, Jupiter, is in Scorpio in the Twelfth House. Keeping in mind what has already been said about Pluto in the Ninth House relative to the South Node in the Seventh, how would Mercury in Pisces facilitate the mode of operation implied in the Seventh House South Node in Gemini?

It would imply that the individual's "teachers" were his or her parents in the early environment, and that the individual was extremely sensitive or impressionable to the impact of the parents' belief systems. For self-consistency and security reasons (Fourth House relative to the Ninth House Pluto) the individual may have drawn partners

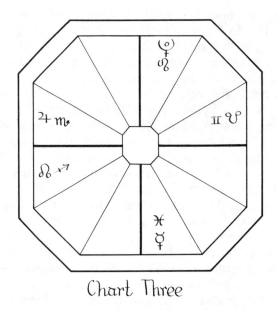

Chart Three

(Seventh House) who reflected the parents' belief system — Mercury in Pisces. The individual did not know what else to believe or how to think independently, so he or she took on the parents' beliefs because they reflected the individual's own desire to understand him/herself in a cosmological context. The individual may in turn pass along these teachings to his or her own children. With the South Node in Gemini, the individual will normally have a tremendous thirst for different ideas and experiences. Yet, in this case, the Ninth House Pluto could have created a situation in which the individual denied the relevancy or legitimacy of any other point of view but his or her own, these being an extension of the parents' views, and later on, a partner's or mate's beliefs. The threat of other points of view would undermine the security structures and self-image — Fourth House Mercury in Pisces.

Yet, with the South Node in Gemini in the Seventh House, the individual would naturally attract others who did have other points of view. So although the individual would attempt denial, he or she would take in this information on a subconscious level. There would be an awareness of something more than the beliefs taken on from the parents and mate. This awareness, from the prior-life point of view, 'sets up' this life. The North Node is in the First House in Sagittarius, the polarity point of Pluto is in the Third House. The planet ruling the North Node is in Scorpio in the Twelfth House. Simply stated, the individual in this life would most likely choose parents who attempted to lay their beliefs upon the child. The child progressively would feel suffocated by those beliefs, and would not feel that the parents truly understood their individual needs. The child would pro-gressively withdraw (Mercury in Pisces) into him/herself in order to begin the lesson of independence (North Node in the First House in Sagittarius, Jupiter in the Twelfth House in Scorpio). By emotionally withdrawing, the individual would also be learning internal security (Mercury in Fourth House Pisces through First House polarity of North Node in Sagittarius). Progressively, as the individual matured, a personal vision, voice and independence would develop which would lead to an evolutionary consequence that would allow the individual to answer and ask his or her own questions. Through philosophical or intellectual confrontation (initially with the parents) the person would not only learn the relativity of beliefs, but also, through meditation (Jupiter in Scorpio in the Twelfth House), would learn the essential unity of all cosmological, metaphysical or religious systems. The Twelfth House Jupiter in Scorpio would progressively dissolve all the old barriers separating one system from another showing the essential unity of all of them. With the North Node in Sagittarius, South Node in Gemini, the individual would see and experience this truth everywhere. The person would also commit to one system that he or she felt most drawn too intuitively because of a natural resonance with it. With Jupiter in Scorpio in the Twelfth House, this system would have to be experienced. It would have to offer techniques and methods through which the individual could directly experience the truth of the conceptual framework being utilized by the metaphysical/spiritual system. In this way the individual would learn to discriminate true teachings and teachers from false teachings and teachers — truth or revelation versus delusion or fic-tion. The individual would learn how to relate with others (South Node in Gemini, Pluto polarity point in the Third House) at whatever level of reality they were coming from and would give to them that which they needed. The individual would be in a partnership that was equal and that encouraged freedom and independence, yet valued commitment. This transformed individual would now encourage independence and respect the individuality of his or her own children.

These basic principles correlating the natal position of Pluto with its polarity point, the North and South Nodes as modes of operation, the planetary rulers of the Nodes as facilitators to the modes of operation, all constitute the main karmic/evolutionary dynamic in the birth chart. This dynamic will serve as the bottom line, the foundation that gives meaning to every other factor in the birth chart. Every other factor in the chart can be related to this dynamic and given new meaning relative to their contributing evolutionary roles implied in this main evolutionary/karmic dynamic.

PLUTO IN ASPECT TO THE NODAL AXIS

When we find Pluto in direct aspect to the nodal axis of the Moon in the birth chart, specific and unique evolutionary and karmic factors exist. The nature of the aspect determines what those factors are.

Pluto Conjunct the South Node indicates one of three possible conditions:

1. The individual is in an evolutionary and karmic reliving condition because of a failure to deal with, or resolve successfully, the issues described by the house and sign that Pluto and the South Node fall in. The planetary ruler of the South Node, by sign and house position, supplies an additional information concerning this issue and necessity.

2. The individual is in an evolutionary and karmic fruition condition. In the past so much effort has been put into the area in question, and with such pure intentions, that the individual is reaping that which was sown before. The individual has some kind of special "destiny" to fulfill. Look to the planet ruling the South Node, its sign and house locality, to supply additional information concerning this

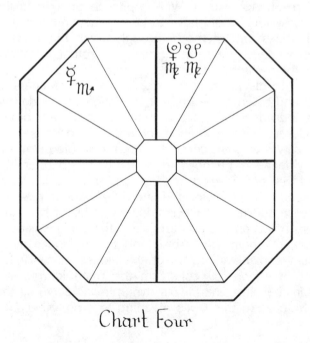

Chart Four

condition.

3. The individual is in a simultaneous evolutionary and karmic condition in which certain conditions from the past must be relived while other conditions are in a fruition state. The house and sign positions of Pluto and the South Node will describe these issues and conditions. Check the locality, by sign and house, of the planet ruling the South Node for additional information.

The first two conditions are rare. The third condition is the most common when Pluto is conjunct the South Node. How can we know which condition exists for that individual? In general, I would suggest you use the traditional astrological approach; observe or ask questions about the individual's life. This technique will allow for a rather quick understanding as to what conditions exists. It can be difficult to understand what condition exists for the individual by simply looking at the birth chart alone.

If you wish to ask questions about the individual's life, then these questions should revolve around the areas and experiences suggested by the polarity point of Pluto and the North Node. If you know the individual, then your observations as to what condition exists should also revolve around these dynamics. If there seems to be a total blockage from being able to realize these polarity points, then condition one exists. If there seems to be something unique and special about the individual's life relative to the experiences and area (house) that the Pluto/South Node symbol is in, then condition two exists. If there are elements of both, then condition three exists.

On the other hand there are conditions in the birth chart that can help us understand the probability of karmic conditions. In general,nonstressful aspects to Pluto tend to indicate a fruition condition. Stressful aspects tend to indicate a relive condition. A combination of both kinds of aspects tends to indicate a dual condition. The nonstressful aspects formed by other planets would indicate, by sign and house locality, those conditions in a state of fruition. The stressful aspects formed by other planets would indicate, by sign and house locality, those dynamics in a relive condition.

In addition to the above, there are other contributing factors and principles to consider when Pluto is conjunct the South Node. In all three conditions, unless there are mitigating factors, the individual will be blocked from being able to realize fully the evolutionary issues described by the North Node until the second Saturn return. As stated earlier, the first Saturn return symbolizes the normal time cycle in which we fulfill, or live out, the evolutionary and karmic conditions of the past. The mitigating factors that can reduce the amount of time spent fulfilling these past conditions are (1) planets conjunct the North Node, or (2) planets in some kind of aspect to the North Node. With a planet or planets conjuncting the North Node, a situation exists wherein that planet has *directly* acted to evolve the individual out of past conditions in the last few lifetimes, sometimes the life just before this one. The specific nature of the planet or planets describes how this was done. With a planet in any other kind of aspect to the North Node, a situation exists wherein it *obliquely* acted to evolve the individual out of past conditions in the laqst few lifetimes, sometimes the very last life before this one. The number of aspects to the North Node relatively determines the reduction of time spent in fulfilling past conditions.

If other planets are conjunct the South Node with Pluto, then those functions (planets) are not only *directly* linked to the past, but are subject to the three possible evolutionary/karmic conditions mentioned above. Apply these same principles to these planets to help determine their condition. If the planetary ruler of the North Node is conjunct Pluto and the South Node, conditions pertaining to the past are intensified two-fold.

Check aspects to the North Node itself to determine the time necessary for fulfilling these past conditions. If a planet is square the nodal axis and Pluto, that planetary function (by sign and house position) is interwoven among issues of the past and future. The most common behavioral manifestation is a situation wherein the individual attempted to evade or escape those issues pertaining to the past. The evasion or escape was linked to the intense degree of conflict or

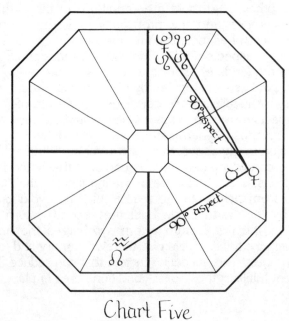

Chart Five

tension that the individual experienced relative to the issues implied by the house and sign position of Pluto conjunct the South Node, and the house and sign position of the squaring planet. While avoiding the South Node problem, the individual attempted to solve problems related to the issues indicated by the house and sign position of the North Node. In so doing, the individual 'skipped steps'. The individual did not succeed in resolving the issues pertaining to the South Node and Pluto within the context of their house and sign positions.

In this life the individual must recover or relive those skipped steps. Only then will the full promise of the North Node be allowed full actualization. Until then the individual is torn in two simultaneous directions — at times manifesting behavior associated with the South Node and Pluto, and other times behavior associated with the North Node. The solution, again, is to fulfill those issues pertaining to the past within the context of the South Node conjunct Pluto: then, and only then, will the promise of the North Node be realized.

In general, the number of aspects to the North Node will indicate the time necessary for fulfilling or reliving prior karmic conditions. In some cases this need is fulfilled at the time of the first Saturn return. This time factor is significantly decreased when the aspects are nonstressful in nature, for they imply a general understanding and partial resolution of these karmic issues in prior lives. A preponderance of stressful aspects tends to imply a lack of understanding or resolution of these karmic issues coming into this life. Thus, the time spent in reliving these conditions in this life is significantly increased relative to the first Saturn return. However, no hard and fast rules can be applied because this evolutionary and karmic analysis must take into account the natural evolutionary condition or station of the individual. The four natural evolutionary states of mankind will be discussed later in this chapter. In general, however, an individual's progression along the evolutionary scale will correlate to a decrease in the amount of time necessary for fulfilling prior karmic issues.

Pluto conjuncting the North Node indicates one evolutionary or karmic condition that applies to all charts. The individual has been working to transform the area represented by the house and sign locality of conjunction within the last few lifetimes, and is meant to continue in that direction. The results of the evolutionary transformation can create tremendous growth in this life. Every other contributing factor in the birth chart will be channeled or focused through the North Node conjunct Pluto. The principle of Pluto's polarity point does not apply in this particular condition.

However, there are mitigating factors that we need to consider. If a *planet* conjuncts or aspects the South Node, then the same three conditions under the discussion of *Pluto* conjunct the South Node can apply. Evaluate the stressful and nonstressful aspects to determine the probable condition of the planet. If that planet is in a fruition condition, then the specific nature of the planet, its sign and house location and the aspects to it from other planets, and their own house and sign locations, will contribute in a positive and integrated way toward the fulfillment of the evolutionary intent described by Pluto's conjunction to the North Node.

If a planet conjunct the South Node is in an apparent relive situation, then the specific nature of that planet, and the house and sign locality that it is in, will act as a conflicting force that to some degree blocks the individual's ability to fulfill the evolutionary impulse of Pluto's conjunction to the North Node. The key to resolving this condition is for the individual to resist the desire or temptation to avoid this factor and its impact on his or her life. Instead, the individual should face those issues head on and integrate and resolve them in the context of Pluto's conjunction to the North Node.

If there are planets in both conditions simultaneously, determine which planet forms nonstressful aspects to the South Node, and which form stressful aspects. The planets forming nonstressful aspects

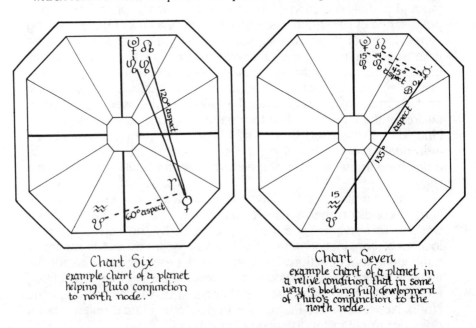

Chart Six
example chart of a planet helping Pluto conjunction to north node.

Chart Seven
example chart of a planet in a relive condition that in some way is blocking full development of Pluto's conjunction to the north node.

will contribute in a positive and integrated way to the fulfillment of the evolutionary intents described by Pluto's conjunction to the North Node, and the planets forming stressful aspects to the South Node must be faced head on and resolved within the context of Pluto's conjunction to the North Node.

With Pluto square the nodal axis, the individual is in a unique evolutionary situation. The individual is equally torn between issues pertaining to the past and issues pertaining to the future. This division exists in each moment in the individual's life. This karmic/ evolutionary condition indicates that the individual has neither succeeded in totally resolving or learning the issues pertaining to his or her evolutionary past (desires), nor the issues reflected through the North Node. Because Pluto is square both Nodes the issues pertaining to each have been acted upon prior to this life. Yet neither has been totally developed, understood, resolved, or integrated. The individual is upon an extremely important evolutionary threshold. The choices that the individual makes relative to his or her desires are extremely critical, as they concern the evolutionary journey and progression. The individual is simultaneously attracted and repulsed by the issues, orientations and lessons of both Nodes. The issue of 'skipping steps' described before is magnified two-fold. The need to

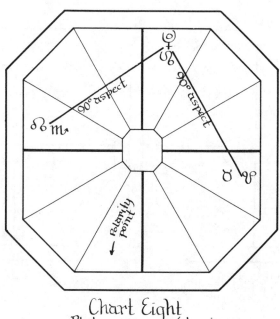

Chart Eight
Pluto square nodal axis

relive and resolve those skipped steps in order to integrate the issues with respect to the Nodes is magnified two-fold. The skipped steps apply in both areas. Until the individual consciously understands the lessons and how to approach and develop them, the behavior will alternately express the orientations implied by the South and North Nodes.

The polarity point of Pluto applies in this situation. This polarity point must be activated to begin the integration and resolution of this evolutionary condition. The mitigating factors to check for are these:

1. If Pluto applies to the South Node, then the polarity point of Pluto and the North Node, with its planetary ruler, must be integrated through the South Node. The individual will have a consistent "bottom line" upon which the evolutionary intentions described by Pluto's polarity point and the North Node and its planetary ruler, can be continually referred. Thus the South Node, and its planetary ruler, will transmute into new levels of expression.

2. When Pluto applies to the North Node, then the polarity point of Pluto and the South Node with its planetary ruler must be integrated through the North Node, with its ruling planet facilitating the process. In the same way, the individual will have a bottom line upon which the evolutionary intentions described by Pluto's polarity point and the

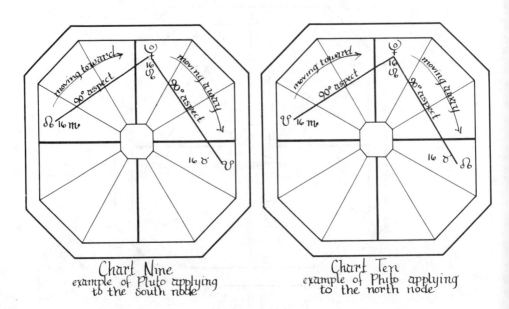

Chart Nine
example of Pluto applying to the south node

Chart Ten
example of Pluto applying to the north node

South Node and its planetary ruler, can be continually referred. The North Node, and its planetary ruler, can then transmute into new levels of expression.

In the case of other planets conjuncting or forming aspects to the nodal axis, apply the same techniques to determine their evolutionary condition. Determining the Node that Pluto is applying to will explain how to integrate and handle those issues. In determining this keep in mind that the mean motion of the Nodes is retrograde. Thus the normal rules for determining what is applying to what are reversed. As an example, let's say Pluto is in Leo at sixteen degrees, the South Node is in Taurus and North Node is in Scorpio at sixteen degrees respectively. Due to the retrograde motion of the Nodes, the North Node is moving toward Pluto. The South Node is moving away from Pluto. Thus the North Node is applying to Pluto and Pluto is applying to the South Node. In this example, then, the integration point would be the South Node. The Node that last formed a conjunction to Pluto is the Node that Pluto is applying to.

PLANETS IN ASPECT TO THE NODAL AXIS
When planets aspect the nodal axis and Pluto has no direct aspect to the Nodes, then those planets have played a major role (by house, sign and aspect to other planetary functions) in shaping the kinds of experiences that the person has had for evolutionary and karmic reasons. The type of aspect determines how the individual responded to those experiences and will supply clues as to the probable evolutionary and karmic condition of those planets. At the end of Chapter Two some chart examples will be used to unite and synthesize all these points.

PLUTO IN ASPECT TO OTHER PLANETS
Any planet in aspect to Pluto indicates that those planets have been, and continue to be, subject to an intensified and accelerated evolutionary metamorphosis. The type of aspect determines the intensity of the evolutionary necessity to metamorphose that planetary (behavioral) function. The stressful aspects produce tremendous evolutionary intensity leading to cyclic cataclysms and restful states. The nonstressful aspects produce a noncataclysmic evolutionary process of a relatively smooth yet continuous nature. When in the middle of a cataclysmic change, the individual will probably not understand why it is occurring. The "why" will be understood, in most cases, after the change has occurred. The nonstressful aspects promote an understanding as to why the changes are occurring during the time frame that the changes are taking place.

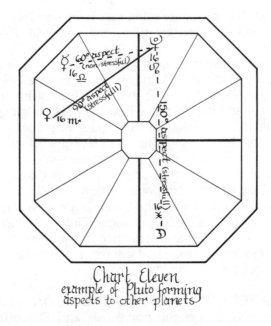

Chart Eleven
example of Pluto forming
aspects to other planets

The number of aspects Pluto forms to other planets determines the degree of evolutionary pace or change in any life. In other words, the number of aspects involving Pluto correlates with the intensity of the evolutionary state; how much they are trying to accomplish or work on in this life. The greater number of aspects, the more the individual is desiring to accelerate his or her evolutionary growth. This principle is clearly a relative phenomenon. Observe the lives of people who have many aspects to Pluto in their birth charts. Contrast this observation with those who do not. The lives of those who have more aspects will be characterized by more cyclic cataclysms than those have fewer. The reason is that these individuals are desiring to confront, eliminate and transmute limitations, stagnations and the status quo at key junctures in their lives. This process is reflective of the desire to return to that which created the Soul. Of course, the individual may not consciously identify this process as such. A few will understand it in this way, but most will not. Simply stated, the individual is desiring to get on with it. People with fewer aspects to Pluto are taking a break from the intensity of evolutionary and karmic necessities.

It makes no difference whether these aspects are stressful or nonstressful in terms of the evolutionary process. On the other hand, the individual's ability to handle and understand this process is governed by the traditional attributes of the hard or soft aspects.

The Sun in the birth chart represents how this evolutionary process is integrated and given meaning. The Moon represents how this purpose, meaning and karmic/evolutionary condition is applied and actualized on a daily basis. This symbolism should be clear when we consider Pluto to be the outermost planet in the solar system, and the Sun to be at the very center. Pluto represents the transmigration of the Soul, and the changing natures of the personalities that it manifests. The Sun gives each new personality of the Soul a way to integrate and give meaning to its life. Since this drama is taking place on Earth, the Moon — because of its proximity to Earth — shows how the new purpose and lesson is applied on a daily, rhythmic basis. The Moon can be correlated to the ego and, on that basis, gives the personality a self-image in order for it (the egocentric personality of the Soul) to feel secure and to know itself.

PLUTO AND THE FOUR
NATURAL EVOLUTIONARY CONDITIONS

Another important mitigating factor to consider are the four natural evolutionary conditions. We must determine the evolutionary condition of an individual in order to understand at what level they are operating. The natural evolutionary conditions will not be revealed by simply looking at the birth chart. Any number of people can be born at the same time and place and have similar charts. But each will be expressing their own state of Soul development and using the natal chart potential in different ways. Not every person with a Ninth House Pluto, as an example, will be a metaphysical or philosophical master. That placement could translate into a master car salesperson who follows traditional religious dictates. We must interact in some way with the individual in order to evaluate and determine what evolutionary condition he or she is within. The evolutionary condition will determine the behavioral manifestations and responses to the evolutionary necessities of Pluto and the nodal axis. This principle of the four natural evolutionary states has its origins in ancient India. The following classifications and terminology have been updated to reflect modern civilization and current sociological theory. These four conditions within the context of the modern world are empirically observable by anyone. The four states are as follows:

1. Dimly evolved or de-evolved state: Two or three percent of the human race is characterized by those people who are either just evolving into human consciousness from other kingdoms (such as the animal state), or are in a de-evolved evolutionary condition due to prior-life karmic causes. Those just evolving into human con-

sciousness have a very dim sense of personal awareness, a basic sense of occupying time and space. These people will appear to be relatively stupefied. In some cases this condition will be associated with mental deficiencies, cretinism or retardation. In other cases mental deficiencies can be found in individuals who, because of past actions in other lives, are de-evolving for a lifetime or two. When one becomes adept at understanding the evolutionary and karmic causes implicated in the birth chart, the reasons for this situation will become apparent. The reasons or karmic causes are different in each case.

2. The herd state: Seventy-five percent of the human race exists in this condition, which is characterized by individuals whose identities are a mere extension of societal norms, beliefs, customs and taboos. In fact, this condition is the mainstream of society itself. It is society. These people do not seriously question what they are told to believe or think. And, if they do question, the answer will be a consensus answer as prescribed by society. In other words, if astronomers tell us that astrology is bogus, then these people will say astrology is bogus without thinking about it on an independent basis. Those who have spent a long time in this state will develop the capacity to lead it.

3. The individuated state: Approximately twenty percent of the human race exists in this condition, which is characterized by those people who question the beliefs, customs, norms and taboos of society. These people seek to discover their own individuality as distinct from society. They desire to know and act upon their own natural laws, beliefs, values, needs, customs and taboos. And they desire to discover these things from within themselves. A vibration of independence will permeate these individuals as they value the right of self-discovery and freedom.

4. The spiritual state: Two or three percent of human beings will attempt to understand their own life and others' lives in a universal/holistic context. In fact, these people desire to understand the nature of all Creation in this context and commonly link themselves to spiritual ideas or teachings as the guiding principles in life. They are attempting to discover, and therefore align themselves with, timeless values, beliefs and truths that apply at all times. In its highest condition, this state produces what are called avatars or spiritual masters and teachers: Jesus, Buddha, Lao-Tzu, Mohammed, Moses, and so forth. This state represents those individuals who have eliminated or are in the process of eliminating all separating or externalizing desires.

These four natural evolutionary conditions are not rigidly delineated classifications. There is movement within and between them. For example, an individual may be at a transitional junction between the

herd state and the individuated state and may express some characteristics of both. When sharing these ideas with another person, or
counseling a client, we must adjust our interpretation and language
to reflect that individual's reality. I cannot overstate the importance of
understanding how these evolutionary factors relate to the birth
chart.

THE THREE REACTIONS TO
THE EVOLUTIONARY IMPULSE

People tend to react to their evolutionary necessities and karmic
requirements in one of three ways. These three possible rections
determine how the evolutionary process will work in our lives. They
are:

1. To resist the evolutionary/karmic requirements altogether.
2. To respond whole heartedly and desire to understand
   what the lessons are for this life and go for it in a totally
   nonresistant way.
3. To be willing to change in some ways, yet resist changing
   other dynamics because of fear of the unknown.

The first response is a reflection of the desire to maintain a
separateness from the Source. The second response is a reflection
of the desire to know the Source. The third response is a combination of both and is the most common choice. We change a little
each lifetime. This is why evolution of the Soul, the progressive
elimination of separating desires, is such a slow process for most.

THE FOUR WAYS PLUTO AFFECTS
EVOLUTION IN OUR LIVES

The four primary ways in which Pluto affects or instigates the
evolutionary intent or process are:

1. By producing emotional shocks in which some behavioral
pattern, or life situation, is forcefully removed from our lives. This
process is associated with cataclysmic change and produces an
evolutionary 'leap'. This situation always occurs when we have resisted the cumulative evolutionary forces to the extent that this effect must
occur in order to enforce the required growth. A classic example is
Nixon and the Watergate affair.

2. By creating a situation in which we form a relationship to
something that we perceive we need. This 'something' can be anything:
a new friend, a new love, a body of knowledge contained in a book
or seminar, the initiation of a new goal, and so forth. The point here
is that by forming a relationship to something that symbolizes that
which we think we need, an evolutionary process occurs. That which

we think we need implies something that we do not already have. By forming a relationship to that 'something' we become that which we need. The psychological process involved here is one of osmosis (Pluto). By forming the relationship, we absorb into ourselves that which we need. A simple example of this process is this: You are now reading this book that you bought because it was relevant or sounded interesting to you. Perhaps the title piqued your curiosity. Maybe you thought you would learn something that you desired to learn. Beyond this book, astrology itself, as a subject, would have represented this to you. By forming a relationship to astrology or this book, through the process of osmosis, you have learned something that you did not know before. By forming the relationship, your own personal limitations, your existing reality (self-definition, self-concept), were encountered. The relationships that we form in this way help us transmute, metamorphose, grow and evolve beyond those existing limitations into new levels of awareness and self-definition. This process is associated with non-cataclysmic change; slow yet progressive evolution.

3. By producing or creating situations in our lives in which we become aware of some external or internal source of stagnation or limitation blocking further growth. This process is quite different than process number two, which involves no particular crisis. It is simple, yet steady, growth. The condition we are dealing with in process number three includes those cycles or times in which we become aware of some internal or external block preventing further growth, and need to find out the source of the block. This process implies a crisis in varying degrees of magnitude. The awareness of the block does not imply or mean that there is an awareness of the origin or source on which it is based. It is precisely this lack of awareness that induces the crisis. This process can have the effect of mobilizing our whole attention in order to focus on the block. This focus of attention can be so intense that everything else in our lives is excluded until we figure out the source of the block. This process is extremely similar to the dormant volcano that suddenly turns active. At some point an eruption will occur that has the effect of transforming the volcano and its surroundings. In the same way, the steady buildup of pressure and the singularity of conscious attention and focus, produces an eruption of the subconscious contents into our subjective consciousness. We become aware, through the knowledge produced in the eruption, of what the block is. As this occurs, we evolve because the source of the block has been removed. We proceed with our life in a new and transformed way because of this cataclysmic evolution.

4. By producing or creating a situation in which we become aware of a new capacity or capability that has been latent or dormant. This process is triggered, at various points in our lives, through internal or external conditions that give birth to some new impulse, idea, thought, or desire. The process of actualizing or developing the new capacity creates an evolutionary or growth effect. A simple example: President Carter became aware of his desire and perceived the capacity to be president when Pluto was transiting his Twelfth House Libra Sun. The process of fulfilling or actualizing that desire and capacity certainly produced growth and personal evolution for him. The process is associated with non-cataclysmic change but requires perserverance and will to actualize the perceived capability.

An important point is that these four processes do not necessarily occur as isolated or separate experiences. One process can trigger or lead to another. As an example, an individual may become aware of a latent capability or capacity at a certain juncture in his or her life. This awareness could then lead to the discovery of a block in a related area that may then lead to a crisis that forces the removal or metamorphosis of that block.

The three possible reactions to the evolutionary and karmic necessities described before are closely connected to these four ways that Pluto affects or instigates our evolutionary requirements. As an example consider an individual who has Pluto aspected to all the other planets and who reacts by resisting growth, change and the evolutionary requirements of his or her life. This reaction could clearly 'set up' the individual for the evolutionary process described in condition one: emotional shocks that lead to the forced removal of behavioral patterns or associations, and as a result, bringing about an evolutionary leap because of the forced removal.

We all have choices to make in our lives. Those choices are reflected in our desires. Desires reflect our evolutionary state or condition, and the reasons for that state or condition. The responsibility for these conditions is our own. There are no 'victims'. Nixon was not a victim. His life situation and experience reflected his own evolutionary condition, needs and lessons. The same is true for all of us.

The assessment of an individual's general evolutionary condition and way of reacting to the evolutionary process must be correlated with the individual's socio/political context. This sociological consideration is important because environmental factors condition how we view ourselves and life in general. To better understand the karmic condition, we should determine why the individual chose the environment, and for what reasons. At the end of Chapter Two this point will be illustrated when we discuss the chart of Adolf Hitler. If you wish to learn more about sociological astrology, I would suggest you read *Wholistic Astrology* by Noel Tyl, available from TIA Publications.

PLUTO RETROGRADE

Now we come to the apparent phenomena of Pluto retrograde. We will consider Pluto retrograde from a purely evolutionary perspective. Pluto is retrograde roughly six out of every twelve months. Approximately half of the population will have a retrograde Pluto in their birth charts. What does it mean?

Relative to the four natural evolutionary conditions, Pluto retrograde means that half the people are not accepting the status quo and will question the status quo in such a way as to reflect their own natural evolutionary condition and state. This process allows for collective evolution at all four evolutionary levels or conditions.

The correlative principle of retrogration can be defined this way: any planetary function (behavior) that is retrograde must be given individual definition. The retrograde principle creates a reactive retreat from the status quo; from societal and collective definitions, expectations, and pressures to respond to life in conforming ways. This reactive retreat allows for individual definition of whatever planetary function is retrograde. In addition, the retrograde phenomenon is a non-static process. It is similar to peeling the layers off an onion to arrive at its core. Thus, there is continual growth (evolution) of the retrograde planet. Pluto retrograde, then, allows for collective evolution to occur because half the people at any one time are questioning the status quo in some ways. This is utterly necessary from an evolutionary point of view because it allows for growth of the entire human species. It prohibits stagnation, non-growth and crystallization.

From the above it may be understood that the retrograde Pluto tends to place the emphasis on the desire to return the Source of the Soul, or to evolve in a more accelerated fashion. Because of the individualizing effect of Pluto retrograde, the person must do that which they must do in their own way. At the deepest possible level (the Soul) Pluto retrograde promotes an internalization process, a relative need to withdraw from external activity. Because Pluto correlates to the deepest reaches of our unconscious, this impulse may not be actively implemented given other factors in an individual's total makeup. The need to withdraw may simply be sensed as a wistful desire that remains unfulfilled. Even if it is not acted upon, the individual will still feel a sense of distance from him or herself and others in a fundamental way.

Because the retrograde Pluto tends to emphasize the desire to return to the Source, or the need to accelerate the elimination of separating type desires, the sense of cyclic or perpetual dissatisfaction is deeper in these individuals than in those who do not have Pluto retrograde. Keep in mind that all of us, with or without Pluto retrograde, will have this experience to one degree or another. Dissatisfaction is directly linked to the interaction of the coexisting desires in the Soul. Until all separating desires are totally eliminated, this sense of dissatisfaction is the psychological symptom or effect that originates in the desire to return to the Source. Dissatisfaction allows for the progressive realization of "not this, not that." Through this process we will someday realize what it is that creates ultimate satisfaction. The point is that those with Pluto retrograde will experience this sense of dissatisfaction more deeply and consistently that those who have Pluto direct. Pluto retrograde accelerates, in its own way, the evolutionary process — individually and, therefore, collectively.

PLUTO — AFTERTHOUGHTS

Earlier it was stated that seventy-five to eighty percent of our behavior is conditioned by unconscious forces that are tied to past patterns in identity association, and that these past patterns were linked to the cumulative evolutionary forces that are rooted in our desires. These past patterns were linked to our unconscious emotional security needs because of the familiarity that they represent; we automatically gravitate to those old patterns because they are what we already are. There is a reliance upon them. The reliance and natural gravitation to these old patterns create blocks to further growth and evolution; we stagnate as a result. The evolutionary force of the Soul induces pressure to eliminate and transform these old patterns into new patterns. The old patterns represent security — the new patterns, insecurity. This transitional process between the old and the new explain many of the apparently negative behavioral characteristics associated with Pluto.

All of us have thoughts, moods, emotions, ways of relating to ourselves and others, ways of understanding life, points of view and so forth. Yet how many of us are aware of how much these dynamics are conditioned by the past? We experience these dynamics 'now'. An idea is an immediate experience and it may be difficult to understand the idea's connection to an old unconscious pattern rooted in the past, and how this past is actually conditioning the present idea itself. And yet it does. This interaction is the very basis of continual evolution; slow yet steady growth. One step leads to the next.

The past has a reason for being, just as the present and future have a reason for being. The past applies to or conditions the present and the future. The transitional process between the past and the future (security vs. insecurity) can create a variety of behavioral 'problems'. Because the present or future (necessary changes, evolution) can be viewed as threatening to our existing security and reality (past), we can resist the future-orientated evolutionary force operating in life. This resistance can create potential dams' that thwart or block the natural evolutionary flow that promotes the state of continual evolution. These dams of resistance create the phenomena of a cataclysmic evolutionary event that has the effect of removing the dam so that the natural flow can continue.

From a psychological point of view these dams of resistance can create complexes, compulsions and obsessions. It can lead to fear, anger, or in extreme cases, rage if the individual experiences emotional shocks or forced removals of something (i.e. a relationship) from his or her life. The individual might be angry and potentially violent because of the feeling that life was 'out of control'. The individual may feel that forces larger than him or herself were in operation. Rarely would the individual understand that the responsibility for this experience originated from within due to evolutionary and karmic necessity. As a result the individual could turn vindictive, mean, or cruel in an attempt to "get back" at whatever caused the emotional pain. The individual could experience a personal hell and create a hell for others as he or she submerged into the deepest recesses of the unconscious emotional and security needs that are rooted in the past (the famous underworld of Pluto). The person who resists the evolutionary pressure or intent to change old patterns might have no perspective or knowledge of why a crisis was occurring — why the partner was leaving. This lack of perspective or knowledge creates the behavioral reactions described here. In other words, the individual would be typically unaware of his or her own personal 'dam' that had finally created or led to a prevention of further growth.

Down the road, after the immediacy of the event was over, not only would there be an evolutionary leap, but also perspective. The individual would regenerate and be "reborn" with a new level of consciousness or awareness. The entire approach to relationships could be radically altered or changed. By plunging into "hell" and experiencing a kind of death, the person evolved into his or her own "heaven" for a time once the perspective gained in hindsight occurred. We all experience relative cycles of heaven and hell with varying degrees of intensity. Pluto symbolizes death and rebirth, metamorphosis, transformation, and regeneration. These processes manifest because of limitations or stagnations that are rooted in our old emotional and security patterns based upon our evolutionary past that condition our present orientations to this life. When these unconscious forces create the "dam" effect, then the above evolutionary and karmic consequences must occur so the individual's evolutionary journey can continue.

Another potential negative Pluto psychological state is guilt or sin. Why and how? Guilt or sin imply a standard of conduct that is "right". Any deviation from this standard would create the sense of guilt or sin. This standard of right conduct is relative and can be linked to a variety of sources. From a spiritual or religious point of view we have the concept of original sin. Beyond the idealized behavior implied in any spiritual system, which can create its own source of guilt, is the idea of separation from the Source. There are many different explanations as to how this separation occurred depending on the specific system. In the Bible this tenet is embodied in the story of Adam and Eve. The sense of separation promotes a feeling of doing something wrong, which promotes a feeling of guilt at an unconscious level. The sense of separation also creates a feeling of having lost something, which in turn can create a psychological "fear of loss", and thus problems of not being able to trust anyone or anything for fear of losing it, or having it taken away. Aligning one's self with a religious or spiritual standard of right conduct automatically creates a structure of internal and external confrontation, which is also a Pluto process. This confrontation makes us aware of our limitations, which are based on what we are not as compared to what we could or should be. This confrontation with our limitations can make us feel guilty, and in some cases angry, as we do not meet our standards, or the standards of the "system" we have aligned ourselves with.

Other sources of the 'standard of right conduct' that can produce the same effects may be ourselves, our parents, teachers, friends, society, lovers, and so forth. It is the relationship between ourselves and whatever source we have formed a relationship with, that can create the sense of guilt when we do not meet the expectations implied in the standard of right conduct.

Another cause or source of guilt is linked to the potential Plutonian phenomena of manipulation. As has been explained, one of the ways that Pluto instigates evolution is through a relationship to something higher or more evolved than ourselves, or through a relationship to something or someone who represents what we need. We can manipulate situations or people in order to get what we need, or we can be manipulated by situations or people so they can get what they need. Manipulation can then create a situation of using or being used, in order to have needs satisfied (Pluto). The intrinsic problem is to maintain the relationship only for the duration of the need, and then to eliminate that relationship because of a new need that dictates a relationship to something or someone else. The sense of guilt is directly linked to knowing that we are using a situation or person for our own reasons or needs. The sense of knowing is, in most cases, subliminal or unconscious. Consequently, this kind of guilt is also unconscious and is held in the memory banks of the Soul. These memories are the result of prior-life actions, as well as a result of potential actions in this life.

It is important to understand that relationships formed in this way are a natural reflection of the desire to grow and evolve; to join, merge, or unite ourselves and our resources with another's resources and state of being (Pluto). Personal limitations are transmuted or metamorphosed because of the relationship formed. Since this is a reflection of the natural evolutionary desire to evolve and grow, the effect of use and manipulation leading to guilt is also natural. This natural effect occurs in order for us to become progressively more aware of our intentions and motivations (Pluto), and to learn from prior experiences in such relationships.

This dynamic has many possible psychological "spin off" effects. One is suspicion, a characteristic Pluto trait. A person who has been used, abused, or manipulated often enough will be suspicious of where someone else is "coming from" — of their motivations and intentions. In very negative cases, suspicion can lead to a fundamental inability to trust anyone or anything, and a natural attempt to contain or conceal oneself because of a fear of being taken advantage of or used. If we are unconsciously projecting this vibration of suspicion we will of course attract that vibration back to us. Others will tend not to trust us. In addition, we would set ourselves up to be misunderstood or misinterpreted by others. This dynamic creates a not uncommon complaint of "nobody understands me". Who is the responsible agent for this situation and why does it occur? It occurs to make us aware of our own inner motivations and intentions. The need to penetrate (Pluto) to the core of ourselves or another in order to find out the "bottom line" (where I or another is coming from and for what reasons) is the evolutionary need.

Another unique Plutonian dynamic is attraction to cultural taboos. Every society or culture has taboos. Individuals with strong Plutonian, Eighth House, or Scorpio signatures can be compulsively attracted to that which is considered a taboo because it represents a potentially powerful experience or association through which the individual may discover a dimension or aspect of him or herself. In addition, that which symbolizes a taboo can be very attractive because the potential to transform existing behavioral limitations can occur through experiencing it. Taboos consequently have tremendous power for some individuals. That which constitutes a taboo can be anything: beliefs, values, norms, customs, a way of doing something and experiences of all kinds. Through experiencing the taboo a growth process will occur even if the experience defined by the taboo is negative in nature.

It is important to understand that an attraction or repulsion to a taboo reflects some very important and basic issues for all of us. One is the issue of control. As an example, what if an individual is attempting to lead a good life, trying to do the inherently right things, make the right choices, yet is still unconsciously "driven" by attraction to a cultural taboo — who is in control? What is in control? What does it mean?

Obviously the controlling force is the Soul wherein these urges, needs and desires emanate or originate. In a universal sense, the intrinsic powers of good and evil are in control. Spiritually speaking, we could say God and Satan. When we link the dynamic of the coexisting desires in the Soul, one for separateness and one for return, within a spiritual or universal context we can now see that the attraction or repulsion to taboos are reflections of these "forces" that are inherently good and evil, positive and negative. If we stubbornly and compulsively manifest separating desires, then the attraction or repulsion to inherently negative taboos can lead into very dark and potentially degenerative behavioral experiences. The origin of these desires is our own Soul, yet, by maintaining or manifesting these separating desires we automatically tap into the largest negative force in the Universe: Satan. The ultimate control then is this Satanic force. Satan's existence is dependent on human delusions and the maintenance of separateness from the Creator: God. These taboo temptations will tend to manifest in areas where we are karmically weak or open to potentially negative or delusive experiences. Pluto's house position and aspects to other planets will describe this potentiality, and where they are likely to manifest.

Conversely, the intrinsic power of good, God, can manifest through our desire to return to the Source. Thus, the cultural taboos that may be encountered in order to foster our evolutionary journey in this direction can also seem compellingly attractive if the culture that we are participating in does not offer the appropriate experiences that will allow for this journey to continue.

Let's utilize a case history to illustrate these principles. This individual has Pluto in Virgo in the Eighth House conjunct Uranus and the North Node. The planetary ruler of the North Node, Mercury,

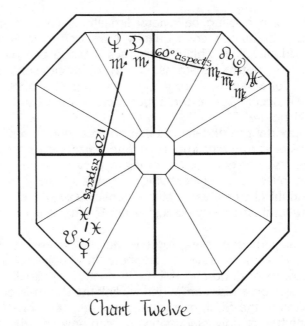

Chart Twelve

is in Pisces conjunct the South Node in the Second House. Pluto and Uranus form sextiles to a Scorpio Moon conjunct Neptune in the Tenth House, and Mercury forms a trine to these same planets. This individual was in the individuated evolutionary state and moving rapidly toward the spiritual state. He was born into a middle class family in Canada. The two primary taboos that symbolized transforma- tive experiences for him were Eastern mysticism and pornography, both considered social taboos in mainstream Canada. The two coex- isting desires in this man's Soul took him into a diametrically oppos- ing directions. The desire to grow and evolve translated into a compulsion that drove him toward Eastern mysticism and the rituals and symbols that allowed for a total reformulation of his personal sense of identity. Yet this desire and direction conflicted with his desire to maintain separateness from the merging of his identity with the Source. In order to maintain his separateness and delusive sense of personal power, he was compulsively possessed with the symbols of pornography. He would frequent movie houses showing por- nographic films, and would buy magazine after magazine with these kinds of photos. In his bedroom, pictures of nude women were pinned next to pictures of spiritual persons. He reported that even as he was meditating, mental pictures of pornographic acts would manifest in his mind. These experiences not only made him feel unworthy, impure and guilty about being on the spiritual path, but also made him feel crazy because he felt out of control. Although he was being tempted by the larger Satanic force, these temptations reflected his own desire to maintain separateness. He was karmically weak with respect to the way his strong emotional and sexual needs and desires manifested. His desire to purify and purge these karmic demons translated into mental pictures, ideas and desires to align himself with a spiritual path that was symbolized by gurus who embodied the purity that he was seeking. The force of God reflected Itself in his life in this way. Each experience and orientation was necessary. Each direction allowed for a metamorphosis to occur. By following the taboo of pornography he got in touch with the basis of his desires and motivations of a separating nature. In other words, by actually following these kinds of experiences the necessary self-knowledge occurred because of them. By contrasting these experiences with the experiences that occurred because of his spiritual direction, he finally became free of the compulsive attraction to this negative taboo. He did this by exercising discrimination and applying his will to cancel the negative temptations when they occurred. In addition, he went on to become a psychological counselor for others who had deviant sexual desires and problems. In this way, he purged the negative desires, temptations and taboos from his Soul.

The taboos that symbolize power and the potential for transformation do not have to be negative in nature. They may be judged as negative by consensus opinion or prevailing social law, however. As an example, Jesus ate with those that he was not supposed to associate with. This symbolic gesture allowed for others to question this "taboo" and thereby be freed from the limitations that this "taboo" implied. Jesus had Pluto conjunct Mars in Virgo, retrograde in the Ninth House in opposition to six planets in Pisces in the Third House. He taught others by opposing the prevailing social custom and "religious" law in this way. Of course he received criticism for this act by the opposing social and religious forces, yet the purity of this act spoke for itself.

It should be evident that even the common behavioral associations that have been correlated to Pluto are based on the natural evolutionary laws or processes that govern life both individually and collectively. The very essence of these evolutionary laws has been related to the coexisting desires inherent in each of our Souls. The interaction of these dual archetypical desires determines that which we think we need in order to grow. This in turn determines the choices that we make (free will), which determines the actions that we initiate. The actions we initiate create reactions (karma), which lead to new actions that are based on the reactions — *ad infinitum*. The coexisting desires are linked to attraction and repulsion, resistance and non-resistance, security and insecurity, death and rebirth; in short, the evolutionary process itself. All psychological problems traditionally associated with Pluto (anger, fear, compulsion, guilt, jealousy, defensiveness, possession, compulsion, obsession, vindictiveness, manipulation, suspicion) have their source in this basic evolutionary law. Conversely, all the positive characteristics of Pluto also have their source in this law — regeneration, rebirth, positive will power, positive motivation of self and others, nondefensiveness, changing as necessary, and so forth. This law is not just a theory, but an observable fact of life itself.

CHAPTER TWO

PLUTO THROUGH
THE HOUSES

In the following chapters Pluto will be discussed through all the houses. Considering all the mitigating factors that were discussed in the previous chapter, including the four natural evolutionary conditions of the Soul, this discussion will present the essential meaning of Pluto in each house. You will have to adjust these meanings to reflect the observed evolutionary condition of any specific individual. In evaluating any house position of Pluto we must also consider the South and North Nodes and their planetary rulers by house and sign. As explained in Chapter One, the South and North Nodes and their planetary rulers, when linked to Pluto, constitute the main karmic/evolutionary dynamic in the birth chart. It is beyond the scope or intention of this book to present or explain all the possible combinations of these factors. However, many examples will be given within the descriptions of Pluto through the houses. In addition, two chart examples will be delineated to help illustrate this process.

Beyond these examples, I would suggest the following procedure so that you can start synthesizing and integrating the main karmic/evolutionary dynamic into your own readings. Each house description of Pluto reflects the archetype of that house. Thus, you can refer to those house descriptions in analyzing the positions of the North and South Nodes. Remember that the South and North Nodes are the modes of operation that have been, and will be, used to fulfill the past and present evolutionary desires of the Soul. These desires are reflected in Pluto's natal house and sign position, and its polarity point. The house positions of the planets that rule the North and South Nodes describe the past and present experiences or vehicles that facilitate the entire evolutionary process.

Ultimately I think we would all agree that the challenge of astrology is to synthesize and link all the differing archetypes, i.e. Pluto in Leo in the First House. The First House is a natural Aries House.

39

Thus, there must be a synthesis of Pluto, Leo and Aries. The point is that each House has its own natural or intrinsic correlation and meaning. Planets placed in a house condition the application, manifestation and orientation of that house. The natural zodiac, i.e. all the houses combined, simply symbolizes the phenomenon of human consciousness. The planets symbolize the specific individual psychology of that consciousness, how it is experienced and put together by the individual. The main karmic/evolutionary dynamic will represent the bottom line or essential emphasis of the psychology of consciousness for any individual, and will tell us the reasons why from an evolutionary and karmic point of view.

Pluto in any sign is a generational phenomenon. Whole groupings of individuals will manifest together to effect the necessary evolutionary requirements from a collective point of view. Each generation must effect the necessary collective evolution through developing the psychological orientation of its opposite sign. In so doing, the intrinsic psychological orientation of the sign that Pluto is in will evolve into new levels of expression relative to the collective evolutionary requirements. Again, it is no coincidence that the Pluto in Leo generation will progressively come into positions of social power and influence in the years to come. The fact that we are rapidly approaching the Aquarius Age reflects the principle of whole groupings of individuals coming together to effect the necessary evolutionary requirements of this age. Thus, Pluto in any sign correlates to the general evolutionary requirements of all those who are born with Pluto in that sign. The specific house placement of Pluto within its sign, correlates to the individualized patterns in prior identity association to which the person will naturally gravitate for emotional security. The polarity point of Pluto with respect to its natal house and sign position, will relate to the specific and individualized evolutionary requirements and necessities that the individual must develop to effect a metamorphosis of these pre-existing patterns in identity. In so doing, the individual will effect personal evolution so that his or her own journey can continue.

PLUTO
THROUGH THE HOUSES
AND SIGNS

PLUTO IN THE FIRST HOUSE OR ARIES

Pluto in any angular house or sign shows that a new evolutionary cycle is beginning. This means that a whole evolutionary cycle has come to a close. Pluto in the First House or Aries will thus correlate to an individual who will feel that they have a special destiny to fulfill. They will feel this sense of special destiny on a very instinctual basis. This sense is based on the fact that there is a new evolutionary cycle beginning for them. The first tentative steps to develop and actualize this new cycle have begun in the most recent prior lives for these individuals.

Since these people have begun a new evolutionary cycle in the very recent past, they will need, and gravitate to, independence and freedom in this life. These individuals will likely be self-centered and somewhat narcissistic in early life. Primary allegiance and identity is with themselves. They have, and will desire, freedom and independence to initiate and fulfill any desire or experience they deem necessary, because experience is the vehicle through which they discover or become who and what they are. This emphasized need for freedom and independence is compulsive, because the new evolutionary cycle that they have embarked upon is so instinctual. There is an intense sense of personal self-discovery that is felt at every moment. In other words, something new could be discovered or realized at any time. Because this sense of self-discovery is not conceptually formulated, self-discovery is linked to that which is to come. Thus, the sense of a special destiny that is waiting to be discovered demands essential independence and freedom. These individuals must be free enough to pursue whatever experience comes their way in order to realize or discover something about themselves that they did not know before. It is the instinctual attraction to the immediacy of a potential experience that serves as the vehicle through which this sense of special destiny and self-discovery can be realized.

The evolutionary intent has been, and is, to find out something new and unique about oneself. This evolutionary pressure sets in motion a process of continual becoming that is highly magnified or intensified. The First House and Aries, by their very nature, demand unchecked expression, action and application. Yet Pluto, by *its* very nature, seeks security through familiarity. Pluto needs perspective and wants to know 'why' prior to the initiation of an action or desire. When these two processes are linked a natural conflict occurs. On the one hand, these individuals naturally gravitate to independence and freedom. They claim the right to be free to do what they must do. On the other hand, they can resist the instinctual impulses to

43

evolve beyond what they are at any moment in time.

The result, on a cyclic basis, is an identity crisis experienced in varying degrees of magnitude. Loss of perspective can occur whenever these individuals cyclically implode upon themselves. Implosions result when the evolutionary desire and pressure to forever move forward through one experience after another, meets the intrinsic resistance of Pluto to secure and maintain that which one already is — familiarity and the known. The identity crisis that occurs is thus based on the simultaneous sense of stagnation, limitation and restriction in terms of what the individual already is, and the need to eliminate these restrictions by initiating new experiences or actions.

The need for new experiences will have the effect of creating insecurity with respect to what the individual already is and how he or she has structured his or her life. Again, these instinctual evolutionary desires to grow, evolve and move forward do not manifest in well-formulated thoughts, concepts, or ideas about what to do. They simply manifest as an intense instinctual impulse to do something, anything, that will create the necessary movement that allows for a continuation of self-discovery. The natural resistance of Pluto when linked to the instinctual impulses to change can result in feelings of inner frustration and anger that can be very intense. Often this anger is projected upon others who are seen as the source of the restriction, stagnation, or problem. In extreme cases the potential for physical violence exists.

The polarity point of Pluto is the Seventh House or Libra. The evolutionary desire or intent of this sign or house is ignited as the cycles of identity crisis continue and deepen. It is ignited as the loss of perspective creates cycles of inner torment and anger. The evolutionary desire will progressively lead these individuals into relationships with others. The process of self-discovery and the actualization of a special destiny cannot occur by maintaining a primary allegiance to just themselves. The prior-life instinctual impulse to maintain freedom and independence promotes a 'natural loner' coming into this life. The vacuum of individual isolation from others that has maintained this prior evolutionary need and impulse, will promote the identity crisis because the person has reached an evolutionary limitation that is rooted in themselves.

The loss of perspective experienced during the identity crisis will manifest as a desire to initiate relationships with others who can help the individual answer the question ::Who am I"?, and ::What am I"?. By experiencing a personal limitation wherein these questions cannot be answered completely within themselves, the individual must learn how to open up to other people. "Who am I?", "What am I?"

will be the stated or unstated cries leading to interaction with others. These individuals will subconsciously attract other people who can have a rather shocking effect upon them. This shock effect can occur through the information, advice, or perspectives given to them by others. The vibrational resonance emanating from their Souls attracts those who will help the individual answer these questions.

As a result of this natural process, those with Pluto in the First House will be instinctively attracted or repelled by many other people. Often the attractions are linked to a sexual or a physical magnetism. Such attractions may or may not be acted upon, depending upon the contributing factors in the person's total nature, Mars may be in Capricorn for example. In all cases, the basis of an attraction to another will be what I call hypnotic attractions that occur on a very instinctual basis: the individual does not consciously know what the basis of the attraction is — it just is. And, of course, others can be attracted to the individual in the same way. The basis of the hypnotic effect, either way, is a natural vibrational resonance emanating from the Soul that creates in each person a feeling that they would learn something about themselves because of the relationship formed to the other. These First House Pluto people naturally draw others who can help them fulfill their evolutionary requirements, and others are drawn to them for the same reasons.

This process works very instinctively and can be quite fleeting in effect. The time duration of this type of relationship does not matter. That which occurs through the encounter does. Some relationships will be longer lasting than others. Some will be sexualized, others will not.

When these individuals feel the instinctual attraction to another, there is a natural desire to find out what the basis of the attraction is. Frustration can result if existing circumstances in life do not allow them to follow through upon the attraction. These circumstances could involve many things including involvement with someone else.

When these individuals feel instinctively repelled by others, they may almost totally deny that person's existence. Sometimes this attraction/repulsion dynamic works in such a way that they feel repelled by the very people they feel attracted to, because the attrac-tion may directly threaten the existing nature of their reality.

Relative to the three basic reactions to the evolutionary and karmic requirements for this life, these individuals can react in three possible ways to the attractions they feel toward others, and the information coming to them from others:

1. total rejection of the information or attraction if the individual

felt threatened or attacked by someone else, or if they felt another's intentions were suspect.

2. to take in the information or follow the attraction wholeheartedly if they felt the intentions or motivations of another were sincere, pure and worthy.

3. to reject some attractions or information and, at other times, take in some information and/or follow an attraction.

As the life unfolds, First House Pluto individuals also come to realize (desire) that they want and need to be in an intimate relationship on a sustained basis, not just brief and intense encounters. From the standpoint of evolution, these individuals are literally learning how to be in relationships on an equal basis. They are learning how to give rather than take, listen rather than dominate a conversation; learning that they are equals with others and that they are just as important or unimportant as anyone else.

By learning how to give to others first, they will learn that their own needs are answered as a result. By learning how to listen, they will know what to give. By listening and giving, they will learn the evolutionary lesson of equality and relativity.

Commonly, most of these people will experience difficulties in relationships that revolve around confrontations of an emotional, intellectual, or physical nature in the early part of life. Often they must experience the sense of being attacked, or attacking, of being misunderstood or misunderstanding another. These experiences occur to trigger the realization of where they and others are coming from — the motivations, intentions, and the basis of an attraction or repulsion. In other cases, intense confrontations may occur through which these individuals experience the sense of not having their needs met by another, or of another confronting them because their needs are not met by the First House Pluto individual. The reasons for these types of confrontations are the same: to promote lessons of equality, of giving, of listening; to fulfill another's needs before one's own, to promote the awareness of one's intentions, motivations and the awareness of changing needs.

Those with a First House Pluto, or Pluto in Aries, will react in one of three basic ways to the evolutionary intent implied in the above experiences. At different times in life an individual may experience each one of these reactions:

1. to seek out relationships in which they are the dominant force, which leads to an unequal situation. They subconsciously attract this situation because it will ultimately guarantee a separation from that relationship. This type of situation maintains the need for freedom but does not meet the individual's evolutionary needs. The

desire for freedom, thus separation, dictates or creates such a relationship. For the very same reason, these individuals may reject or repel relationships altogether. In this case, they fear becoming overly embroiled in another's demands and needs. If they become overly involved in another's needs, they would consider such situations as potential "detours" that interfere with their own desire to actualize their "special destiny". Some individuals, in this reaction, will pop in and out of relationships as the desire or need dictates. The identity crisis is now whether or not to be in relationships, as cycles of personal limitation and stagnation occur.

2. to seek out relationships in which they are being dominated, which also leads to an unequal situation. This is an extreme reaction to the desire to return to the Source and is usually caused by an old karmic pattern that has led to subconscious guilt. The guilt is usually based on the subliminal knowledge of being overly egocentric and very manipulative in other lives. Thus, there is a need to "atone" for the guilt. This is accomplished by drawing the type of partner who is totally dominating by nature. They dominate by compulsively confronting and pointing out the individual's weakest areas. This type of partner presents him or herself as the individual who will reconstruct or reformulate the First House Pluto person according to his or her idea of how these individuals should be.

3. to seek out relationships that contain the potential for equality, with each partner giving and receiving in equal ways, each partner helping the other to understand themselves in open and non-defensive ways, and each partner allowing for independence in the other. Each would value commitment to the other and the relationship. Each partner would desire to define the relationship in "new ways", and be open enough to let the relationship evolve and change forms as necessary.

The evolutionary prescription demands that these individuals learn how to balance their need for freedom and independence with their need for relationships. Rather than falling into an either/or situation, the evolutionary need is to realize that both needs and desires are relevant and necessary.

Such individuals must learn that the optimal relationships for them are with others who encourage and allow for their freedom and independence. They must learn to recognize that this rhythmic need to be alone or to be with someone is not predictable, nor is it known to the individual prior to its instinctual manifestation. The challenge is to follow these opposing rhythms as they present themselves. If this is not done, then loss of perspective, emotional distortion, misidentifying the cause of a problem, and an identity crisis will occur. If these rhythms are followed, then these reactions will be minimized.

Because the First House Pluto person is in a new evolutionary cycle, the individualizing impulses manifest on a very instinctual basis. These people will not know the reason of these desires, nor how to explain them or account for them in a conceptual sense. As a result, they feel less competent than others at explaining themselves, their actions, their needs and who they are. Negatively, this dilemma can create a situation in which the individual adopts someone else's values and beliefs. This inevitably leads to an extreme reaction in which the individual rejects those beliefs at some point. In this reaction, the individual retreats into isolation in order to recover or discover who they are as an individual, independent of what they have allowed themselves to adopt. Yet, because the evolutionary pressure is to be in a relationship, the person ultimately reacts against this isolation and seeks out relationships. This process will keep repeating until the person realizes that comparing him or herself to others in this way is not productive. Then the person will seek out relationships with others who listen to and encourage their individual self-discovery as it occurs, rather than forming relationships to others who try and dictate who and what they are, or try to make them into something that they cannot be.

Positively, these First House Pluto people manifest the courage and will to be different, to strike out on their own in a leadership capacity and to serve as examples to others that they too can assert their right of individuality in the ways they deem necessary. Their lives can serve to motivate others to do what *they* must do in order to realize their own special destinies.

Let's utilize an example to illustrate these dynamics and principles. This individual has Pluto conjunct the North Node in Cancer in the First House. The planetary ruler of the North Node, the Moon, is in Leo conjunct Neptune in the Third House. The South Node is in Capricorn in the Seventh House and its planetary ruler, Saturn, is in Scorpio in the Fifth House. Pluto is square Venus in Aries in the Eleventh House. Venus also squares the North and South Nodes. This individual is an American and is in the spiritual evolutionary state. Pluto conjunct the North Node in the First House demonstrates that he has been working within this area of evolution before and is meant to continue. In other words, he must maintain primary independence and freedom in order to actualize his special destiny. Yet with Venus squaring Pluto and the nodal axis there is unfinished business with respect to intimate relationships and the emotional issues therein. With the South Node being in Capricorn in the Seventh House with respect to Saturn in Scorpio in the Fifth House, his own approach to

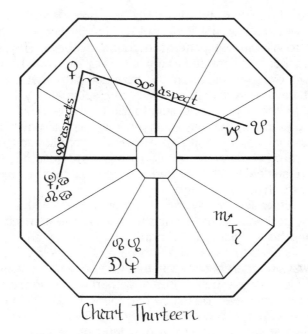

Chart Thirteen

relationships was based on his authoritarian nature. He needed to be the person in control of the relationship in order to feel emotionally secure. These symbols also indicate that in prior lives he had inherited positions of social responsibility and power from his father and he had likely been a member of the nobility and the ruling hierarchy. His marriage partners were also picked from the noble or ruling class. These arranged marriages did not fulfill his emotional needs, and he seems to have controlled these situations by controlling his own emotions. He withheld his emotions through fear of rejection and because he did not know how to express them. This emotional dissatisfaction promoted a rebellion and so he left the constraints of his family and the inherited obligations.

He began to travel far and wide searching for ideas and truths of a transcendent and spiritual nature. His relationships were now formed with teachers and communities of like-minded individuals. He attempted to live his life with respect to intellectual and spiritual ideals, and to go it alone. Wandering through many lifetimes in this way, he began to evolve very rapidly on a spiritual level. Yet, in this present life, he was destined to confront the unfinished business of intimate relationships in order to recover the "skipped steps" of the emotional supressions of many previous lives. His Soul needed to do so because further growth at a spiritual level could no longer occur. In this life he gravitated to all the old patterns mentioned above. He became a spiritual teacher and attracted many followers. He wrote books based on his own experiences upon the spiritual path that encouraged others to take charge of their own lives in a spiritual way. He gave his life totally to these purposes and fulfilled much of his unfinished relationship karma in this way. And yet, it was an ultimately detached and impersonal way of doing so. He would not allow others to get close to him on a personal emotional basis. He became suffocated in the "special identity" that he had as a spiritual teacher. Others saw and held him in a special light.

This suffocation finally translated into yet another rebellion, wherein he desired and threatened to disassociate himself from his position and his special spiritual name. This identity crisis was founded upon his need to relate with himself and others in a much more personal way, in an emotional way that was based on equality. Then, on one of his journeys, he ran into a woman who resonated with his Soul on a personal level. He felt he had found his Soul Mate. She felt the same way in the beginning. They married. He took her back home to live as an example of a new kind of spiritual relationship. She served as a vehicle through which he could recover and feel the emotional dynamics and issues that had been supressed for so long. Yet, in the end, she left him because she was not, in her own evolutionary state, prepared for the role he had carved out for her.

This event also occurred as an element of karmic retribution for his own leavings and rebellions during other lifetimes. But it also promoted a positive outcome. The sudden and unannounced leaving forced him to examine himself in necessary emotional ways —where he had been coming from, and for what reasons. It promoted a metamorphosis wherein he became emotionally free to express his feelings, moods and emotions with those he trusted. He would now hug people, hold their hands, and above all, laugh at himself. The suffocation of his old self-imposed image and identity, reinforced by his followers, now gave way to a more human and well-rounded identity and image that allowed him to receive the emotional input and giving of others. He finally wrote a book on how to spiritualize the dynamics of marriage.

Once these evolutionary lessons have been learned by these First House Pluto people, then their ability to balance their own legitimate needs with the needs of those around them is unsurpassed. Their evolved ability to listen to another will allow them to give to another exactly what they need. In so doing they will attract to themselves others who will have the capacity to give to them that which they need. They have intrinsic courage and capacity to break new ground in whatever aspect of life that they apply themselves to, and can give courage to others to do the same thing. Furthermore, these individuals can become aware of the basis and nature of their desires, and why and how those desires determine their reality. They can apply this knowledge to others as well, and help others become aware of the basis and nature of their own desires. This knowledge can be used to help themselves and others make the right choices as to what desires should be actualized, and which ones should not.

Common characteristics of the First House Pluto types are these: Intense individualists, can be very strong willed, intense, magnetic, bull-headed, defiant against arbitrary authority, courageous, possess inherent leadership abilities, are not given to meaningless conversations, have a penetrating gaze, and can be hard to get to know deeply. They have very strong physical bodies.

Famous people born with Pluto in the First House or Aries:
>Annie Besant
>Ram Dass
>Prince Andrew
>Prince Charles
>Karl Marx

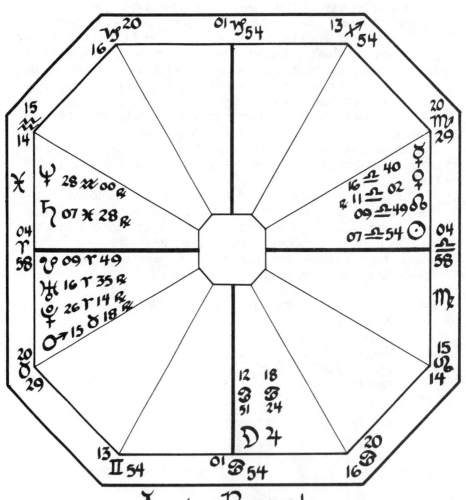

Annie Besant
source: Marc Penfield

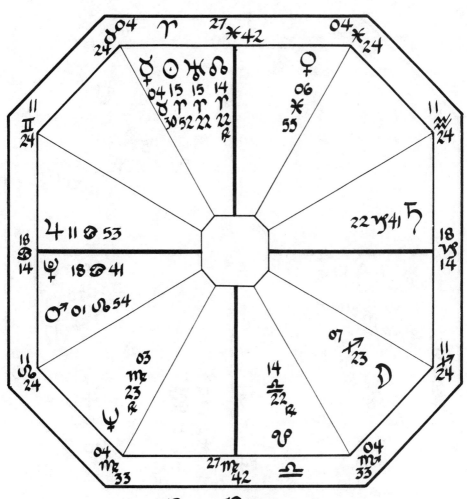

Ram Dass
source: Lois Rodden

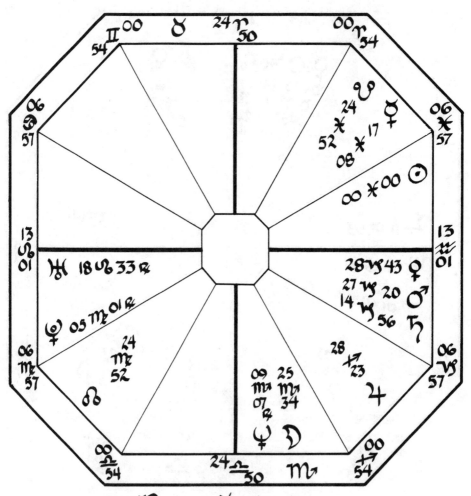

Prince Andrew
source: Lois Rodden

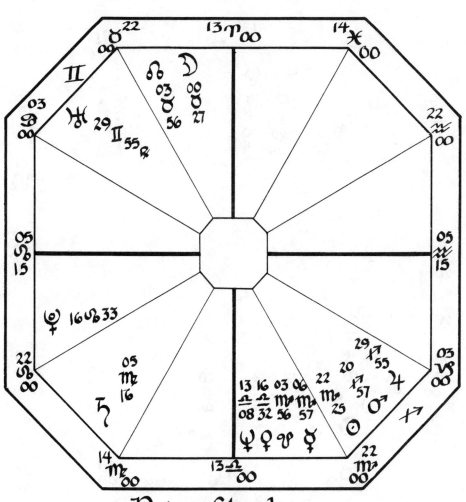

Prince Charles
source: Marc Penfield

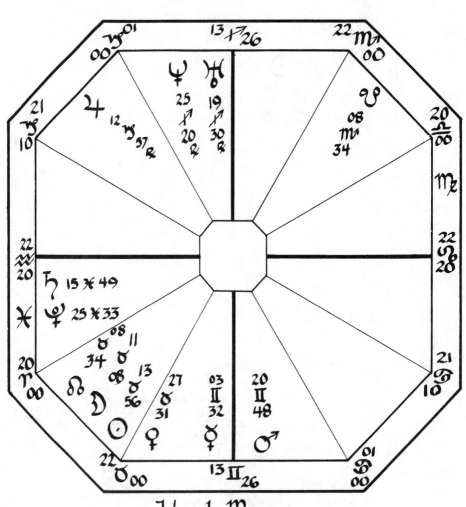

Karl Marx
source: Marc Penfield

PLUTO IN THE SECOND HOUSE OR TAURUS

Pluto in the Second House or Taurus indicates a prior evolutionary desire and need to develop self-reliance and self-sufficiency. In the broadest sense, Taurus and the Second House correlate to physical and biological survival of the human species. Just as the species has learned to survive physically by identifying the resources necessary to accomplish this, such as learning to make fire, so have these individuals learned how to identify their own personal resources to sustain themselves. Those with Pluto in the Second House or Taurus have an extra strong survival instinct as a result. From a biological standpoint, they have a very strong sexual nature because a function of survival is to reproduce the species.

These individuals come into life with an emphasized need to survive and sustain themselves on the physical, and therefore, emotional plane. They will come into this life with a natural self-reliance and natural self-sufficiency. In prior lives they have been learning how to identify their own personal values and needs in order to develop these qualities. Whereas the First House Pluto had to initiate action and experience in order for self-discovery to occur, the Second House Pluto individual has needed to withdraw into him or herself in past lives and will naturally gravitate to experiences of withdrawal in this life. By withdrawing, these individuals have internalized their conscious focus in order to discover their own values and needs from within.

By remaining rooted (fixed) in themselves, these individuals have experienced, sensed, or discovered their own essence as contrasted with the changing flux of circumstances around them. As a result of this process, these individuals commonly come into this life with what I have called "frog in the well karma". In other words, the frog in the well has identified a small piece of the sky — that which it can see from the bottom of the well. Furthermore, the frog has identified this small piece of the sky as the entire universe. The frog, of course, is quite secure and safe in the bottom of the well. In the same way, these individuals come into this life with a limited vision or knowledge about themselves and life in general. This limited vision is known, familiar and secure. Thus, there is a reliance upon this limited vision because it "works". This reliance upon the past can create an inertia or laziness, and a resistance to jumping out of the well. The degree of resistance and limited vision is relative to the mitigating factors determined by the signature of the main karmic/ evolutionary dynamic in each birth chart. As an example, if the South Node is in Gemini the degree of resistance and limited vision would be less severe as contrasted with the South Node in Capricorn.

Relative to physical survival, those individuals in a herd state evolutionary condition tend to identify material values as a vehicle to self-sustainment. We can see the implied limitation here. Others in the individuated evolutionary condition have learned to identify them-selves as the resource creating the means to self-sustainment. Those in the spiritual or universal evolutionary condition have learned that the Source or Universe provides that which they need in order to sustain themselves when one is receptive to that provision. They commonly isolate themselves from the environment and others in order to concentrate upon their relationship to the Source.

In all cases, limitation is implied. For those who have related to material values as a means for self-preservation or self-sustainment, the limitations are the values themselves. The material needs and desires for these individuals can be a bottomless pit because they believe that physical security brings emotional security. Relative to the fear of loss intrinsic to Pluto, there is a subconscious fear of los-ing (or not having enough) material goods. Possessions and money equal power because of the status that is implied in material wealth. Some of these people are not beyond manipulating the possessions, resources, or money of others in order to gain them for themselves. Others are not beyond forming vicarious relationships to situations or people who represent, or are the embodiment of, this kind of power and status. In other words, some of these people have iden-tified that they have these needs and desires, yet are too lazy to make the necessary effort to create the resources for themselves. It is easier not to do so. Instead they form vicarious relationships to others who have such resources and become extensions of their identities and values.

A small percentage of those that manipulate the resources of others to gain those resources for themselves have used very devious or underhanded tactics to do so. A small percentage of those who have made minimal or no efforts to actualize their material needs may live beyond their means; driving the fancy sports car they can-not afford, charging the limit of their credit cards, going into debt. Individuals who have made the necessary effort to actualize their material needs can accumulate vast sums of money and possessions due to the concentration (Pluto) upon the desire and effort.

With the polarity point being the Eighth House or Scorpio, the evolutionary intent will lead these individuals into intense internal and external confrontations. Such confrontations will reveal the limitations implied in their value systems, in how they have learned to relate to themselves, and how they have identified the nature of their personal reality — the frog in the well.

For those who have amassed wealth, the shock of confrontation will involve the realization that material abundance does not equal happiness or emotional security. The Eighth House, and Pluto itself, evokes or symbolizes the awareness of deeper levels of reality; that there is more to life than material values.

Those who used devious means to obtain wealth will be haunted to their graves by a gnawing sense of guilt. Guilt would serve to shine a light upon the internal motivations leading to this situation. This in turn can lead to deep internal shocks of self-knowledge because these individuals have been typically "unconscious" of these deeper motivational forces. In other cases, the wealth may be taken from the individual in order to enforce a re-evaluation of their value system. The issue of self reliance is then experienced in a new way.

Under some evolutionary and karmic conditions, individuals with a Second House Pluto are born into difficult material conditions of lack in order to enforce lessons of self-reliance. In other cases, some individuals will be born into life situations in which there is wealth linked to the parents but they are denied the material resources because they are not able to relate to their parents' values, or can not conform to the dictates of the parents' desires. In a few rare cases, the individual can experience the shock of imprisonment if criminal means were used to acquire the material wealth. In other cases the individual will be made to struggle throughout the entire lifetime for material security. No matter how much effort they put into sustaining themselves, prior karmic causes will dictate that they live or survive on a marginal basis. An overall evaluation of the entire evolutionary/karmic issues pertaining to the individual's past must be made in order to understand the reasons why any or all of these conditions exist in this life.

For those who have been living beyond their means, the Eighth House polarity will at some point produce shocks of repossession because of the compulsive spending patterns. If material goods are repossessed, or if the individual experiences the increasing pressure to keep making money to pay for these material goods (which are a form of bondage) then at some point the internal or external shocks experienced may produce the necessary awareness as to why they are operating in this way. Again, the frog is forced from the well.

For those who have identified themselves as the resource for self-sustainment, i.e. some talent, capacity or ability, then the Eighth House polarity enforces the lesson of merging with another or others. These individuals must open up to other people or situations through ways in which their own personal limitations and ways of relating to reality are exposed. They will have a deep sense of who they are, and they will relate to themselves and others from this more or less fixed sense of self-definition. This attitude not only produces self-limitation, but also personal isolation. These individuals may have identified their essence, their capabilities, their personal values, but if they do not know how to implement or apply them to "forces" outside themselves, i.e. society and other people, then the Eighth House confrontation of limitation, denial or isolation will occur. It may be wonderful to have natural talent as a potter, yet if the effort is not made to integrate, apply, or link the ability to societal needs in ways that are compatible with others' value systems, it will not be possible to sustain oneself with that ability. Evolutionary and karmic necessity will create the effect of denying the individual's ability to sustain him or herself relative to the intrinsic capability that they possess in order to do so.

Individuals with a spiritual focus must learn to open up to new rituals, techniques or methods that allow for a further deepening of their spiritual commitment. They must also learn to share their spiritual resources with others in some way, rather than living in isolation and nurturing only themselves. By opening up and sharing themselves a metamorphosis occurs due to the implied limitations from the past. If they do not learn to open up and share, then the Source or Universal will cease to provide the material needs necessary to sustain themselves. In addition, the small piece of the Universal sky that they have identified from their wells will progressively become filled with the clouds of stagnation. As a result, the deep inner core of spiritual sustenance will dry up. The resulting inner confrontation can produce the necessary metamorphosis through which these individuals open themselves up to their karmic and evolutionary requirements.

In all cases the Eighth House polarity point demands a confrontation with personal limitations. The frog must be forced from the well to expose it to the full light of day. This process occurs by forcing the individual in upon itself in some way to examine the internal dynamics that are creating these life situations. Thus, they are being forced out from the well to examine the larger and deeper forces at work within them: the motivational patterns and the reasons for those motivational patterns, their desires and the reasons for those desires, how they have been relating to themselves and why they have been relating to themselves in that way, what resources they have identified with in order to sustain themselves and why those resources, and what personal values they have identified with and why those particular values. By doing so, these individuals create the potential for a total metamorphosis: to bring about a totally new way of relating to them-selves, others, and reality in general. In this way, these individuals can adopt redefined value systems that allow for the necessary growth to occur. This natural evolutionary process occurs in such a way that these individuals subconsciously draw or attract life situations in which they experience the limitations associated with their value sys-tems, needs, self-image and the ways that they have been relating to themselves and others. The personal implosion that this produces forces the individual to examine all of these dynamics in order to find out why they have identified themselves in a particular manner.

Progressively these individuals will learn to unite their own resour-ces and lives with others so that a personal transformation can occur; a transformation of limitations. Because of the implied inertia of the evolutionary past, changes occur slowly. Most of these individuals need to know "why the change" prior to making the change. This fix-ity creates the potential for internal or external "atom bombs" or cataclysmic upheavals through which the landscape is utterly trans-formed. Because of the evolutionary intent of the Eighth House polarity point those with Pluto in the Second House will attract, at key points in life, intense and powerful people. These people will be timely messengers that deliver or create the necessary encounters and confrontations of personal limitations. These intense encounters will allow the individual to experience him or herself, negatively and positively, in the ways that they are "hung up" or blocked. The poten-tial for growth exists if the individual is willing to transform and jump out of the well of personal limitations.

These key people will have the ability to understand or "see" the source of the limitation that has blocked the individual. By focusing upon it, they can help the Second House Pluto person to grow and evolve, to relate to themselves and others in new ways, and to link themselves to new or redefined values. These key people will promote self-effort and self-reliance in the Second House Pluto person, rather than promoting dependency upon themselves. If and when other powerful people show up in the Second House Pluto person's life who promote dependency upon them, then the individual should not get involved with them in any way. If he or she invests too much power in, or gives too much power to, either type of person, then the Eighth House polarity point of Pluto will guarantee that a forced removal of that person from the individual's life will occur in order to enforce the ongoing lessons in self-reliance and self-sufficiency.

Individuals with Pluto in the Second House have strong physical and sexual natures. Evolution takes place when they learn to merge

and unite with another on an emotional basis, rather than approaching sexual activity in an essentially masturbatory way by using another
for physical/sexual gratification and release. It should be understood
that the emotional power of Pluto in the Second House naturally
leads to strong physical and sexual needs for these people. Again,
the Second House is an archetype that correlates to the instinctual
impulse to procreate the human species. When this instinctual impulse
is linked to self-reliance, the result will be a strong masturbatory
nature in these individuals when they are not involved in a relationship.
The natural emotional build-up of the energy of Pluto in any house
demands cyclic release through that house, or its polarity house, in
order for emotional stability to occur. If it is not cyclically released,
then emotional distortion and loss of perspective can occur in all
aspects of the individual's life. Pluto in the Second House, then,
demands that this emotional build up of energy be released in
physical/sexual ways. Even when these individuals are involved with
another, it is not uncommon for them to still need to masturbate
anyway, because masturbation is a means through which these people
"ground out" with themselves. It is a way to continue the evolutionary
lessons in self-reliance and self-sufficiency. The point here is that this
phenomena is natural for these folks, and it should not be judged
nor subjected to behavioral modification.

Some individuals with Pluto in the Second House have, or will
use, their sexual/emotional power to possess, use, or manipulate
others. They may withhold sexual activity from another, or they may
use their sexual nature to gain something from someone. In extreme
cases this can translate into "notches on the belt" as these types go
through one sexual partner after another. This phenomena occurs
so that these people can prove their emotional and personal power
to themselves. When compulsion and obsession occur relative to
physical/emotional/sexual needs, some of these individuals will become
almost singularly focused upon sexual images, taboos and symbols.
Obviously, this situation implies a limitation. At some point in their
lives, an emotional shock and confrontation will occur to metamorphose
this behavior and emotional/physical orientation.

Individuals who are in the spiritual or universal evolutionary state
must learn various rituals or techniques that use the energy of sexual
union to promote a transcendence of the physical plane. In other
words, to link themselves with the appropriate techniques, methods,
or rituals that teach them how to use the sexual/emotional energy
experienced during sexual activity to expand their consciousness. An
example of these techniques and rituals can be found in the ancient
tradition of Tantric Yoga, which originated in Tibet. The Kama Sutra

of India and Taoist Yoga techniques from China are other traditions that approach sexuality in this way.

Once the Second House Pluto individual begins to effect the necessary evolutionary and karmic requirements of the Eighth House polarity point, the results will translate into an individual who not only understands their limitations, but who will consciously make the effort to unite with the appropriate activities that will promote the required mutation of those limitations. These activities will be different in each case. The evolutionary need and theme will be the same in all cases. As this process unfolds, these people can become very good at pointing out the limitations and why those limitations, in others, and suggest the appropriate means for transformation. These people can learn to merge themselves with others, to merge their resources with others, and show others how to do the same. The transformation that occurs in these merging activities will only add to what the individual already is; that which was brought in from other lifetimes. They can learn how to merge without becoming dependant upon that which they merge with. Thus the ongoing lessons in self-reliance will occur while allowing for a transmutation of the existing limitations brought with them into this life. Evolution has been effected as a result.

Let's utilize a case history to illustrate these principles. This individual has Pluto in the Second House in Gemini. The South Node is in Virgo in the Fifth House and its planetary ruler, Mercury is in Cancer in the Third House. The North Node is in Pisces in the Eleventh House and its planetary ruler, Neptune, is also in the Second House in Gemini. It is not conjunct Pluto. This individual was in a transitional evolutionary condition between the herd and individuated state. He was born into the upper middle class in the United States and was Anglo-Saxon. This individual went to a university and became trained in medical psychiatry. He utilized the resources of his father to do so. He related to his father's value system and beliefs. He had a very strong drive to actualize his destiny; he sensed that his destiny was special. He was very curious about the workings of the human mind and the dynamics that effected it. This curiosity, coupled with his need to serve humanity, drove him to learn about the mind from a psychiatric point of view. He became an excellent psychiatrist and was renowned for his diagnostic abilities. He wrote books, lectured and traveled in order to teach others about his opinions concerning the psychiatric model and the drug therapies that he felt to be right. Of course he sustained himself physically and monetarily on this basis. His major problem, and the basis for the intellectual challenges that he encountered from colleagues was that he was fixed upon his

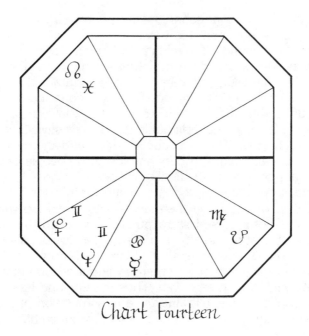

Chart Fourteen

own ideas and opinions. He was certain that they were right and that other points of view that did not agree with his were wrong. Beyond this, the discipline of psychology was irrelevant and misleading according to him. His need for power and recognition were so strong that they blinded him to the potential legitimacy of other disciplines or points of view, which made him feel impotent, inferior and insecure. And of course astrology, parapsychology, and the like were dismissed out of hand. He had opinions about everything whether he had studied them or not. From the well of his psychiatric model, he ordered the entire cosmos to reflect his point of view.

As he became older his ideas became obsolete as others made advances not only within his discipline, but in other disciplines as well. Some of his ideas, once held in high esteem, were seen to be wrong as these advances and breakthroughs were made. He fell out of fashion and his teaching position was lost because he was unable to change and move forward. His books went "out of print". As these personal shocks and confrontations mounted his own intellectual and emotional state began to suffer. Emotionally he became cyclically depressed. All of his possessions and money did not make him feel any better. In fact they made him feel more lonely and empty. As his mind disassociated from his ideas and opinions, new thoughts and ideas seemed to appear in his mind of their own accord. He did not know the origin of these thoughts because they were not based on his deductive thinking or analysis. These new thoughts and ideas directly confronted his most cherished opinions. He finally gave in to this process and began to meditate in his own way. Along with meditation he began to study many subjects that he would not have even considered before. Over a process of many years he utterly transformed and laughed compassionately at others who were just as convinced about their own views as he once was. He professed the idea and philosophy of relativity in his last year, and sold his big expensive house. He lived in a small apartment overlooking the ocean. He died in a chair as he was gazing at this ocean, and there was a smile of peace on his lips. He had made a final leap into the individuated evolutionary state.

Common characteristics of the Second House Pluto individual include: stubbornness, defensiveness, stability and inner strength, "survivors", patience, perserverence or laziness, slow to change, quietly powerful, quietly intense, a need to link themselves to something that has power or is powerful, the need to possess that which they think they own, or that which is "theirs". Conversely, they can promote self-reliance in others, and realize they own nothing.

Famous people with a Second House Pluto:
>Henry Miller
>Adolf Eichmann
>Larry Flynt
>"Cher" Bono
>Hans Christian Andersen

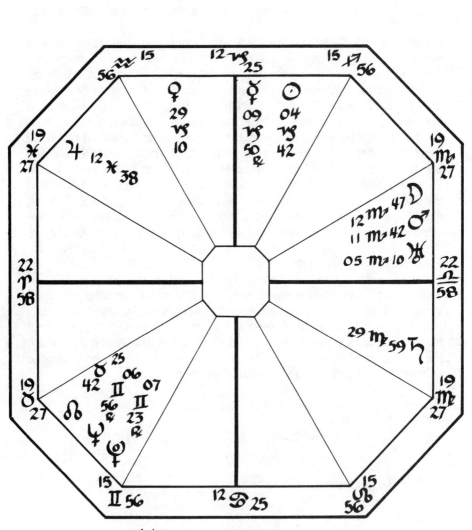

Henry Miller
source: Lois Rodden

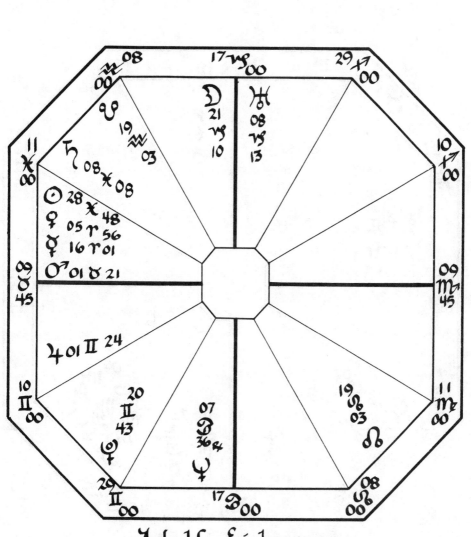

Adolf Eichmann
source: Marc Penfield

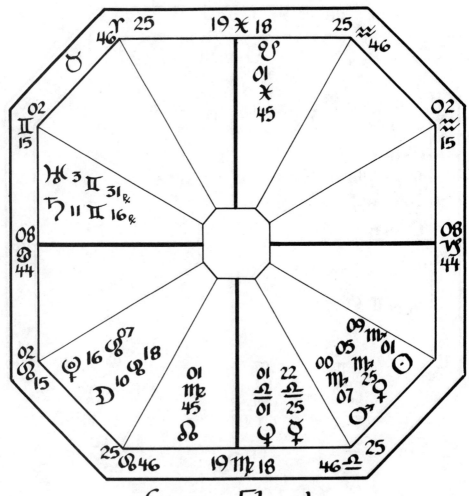

Larry Flynt
source: Lois Rodden

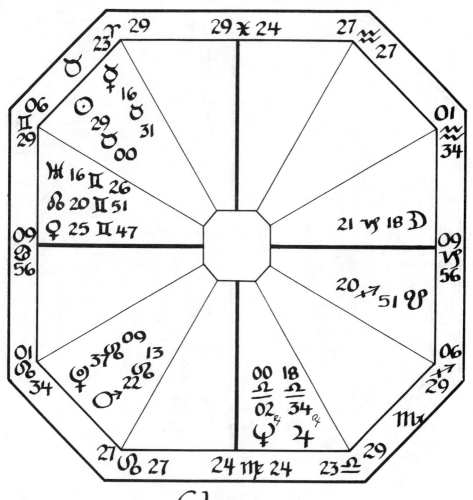

Cher

Source: Marc Penfield

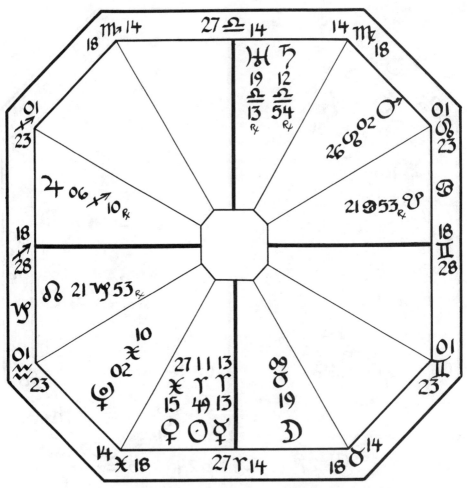

Hans Christian Anderson
Source: Lois Rodden

PLUTO IN THE THIRD HOUSE OR GEMINI

Individuals with Pluto in the Third House or Gemini have experienced the evolutionary necessity to project themselves into the physical environment in order to collect information, facts and data. Thus, this process has led to the evolutionary development and emphasis upon the mind and intellect.

In the Second House, evolutionary necessity required withdrawal and internalization in order to discover the sense of individuality from within; to give it value, meaning, and the ability to identify one's own unique resources to effect self-reliance and self-sustainment. In so doing, emotional stability and security could be realized.

In the Third House, the individual necessarily has had to move out from this center of subjective isolation — the frog in the well. Those with Pluto in the Third House have had to project themselves into their immediate physical environment in order to make broader and larger connections to it, and thus to themselves. These individuals have gained new experiences through which they have expanded their ideas and conceptions of who and what they are in relation to their physical environment. They have been developing their mental powers and intellectual abilities to logically order their existential existence. In other words, they have needed to give names and classifications to the objects and forms of the physical environment in order to understand and know it. By doing so they have also defined themselves by understanding, or attempting to understand, their relationship within the scheme of things — or, as the Taoists would say, "within the ten thousand things".

In the broadest possible sense, the Third House is the need for the human species to give names and classifications to what is otherwise a phenomenal world. By naming and classifying, we have made ourselves intellectually, and therefore emotionally, secure. In the deepest sense, the Third House relates to our need to know the physical laws of our world in order to understand how it works.

Thus, individuals with the Third House Pluto have desired to experience many kinds of circumstances and situations of their own making. Their natural intellectual curiosity has allowed them to build upon, from their own evolutionary point of view, the idea of who they thought they were and think they are now.

There has been a desire and need to understand their "world" in a larger and larger framework. There has been, and will be, a desire to accumulate a storehouse of facts and information about their world in order to build a logical framework or superstructure of ideas that rationally and empirically explain their relationship to the environment. The desire in the past has not been to understand the *deeper meaning* of these facts; the metaphysical or cosmological laws as contrasted with the physical laws of the world. The focus has been, and will be, upon facts which are verifiable through the senses.

The emotional security of individuals with a Third House Pluto has been linked with their ability to logically order, and thus "know", their environment and their relationship to that environment. By knowing their environment these individuals know themselves. This dynamic of emotional security linked to the ability to intellectually organize reality sets up a unique problem. On the one hand, the evolutionary desire and need has been to expand upon the intellec- tual framework. On the other hand, the desire for emotional security has led to the desire and need to create a foundation upon which the expansion is built; a specific point of view. The problem is that the desire to expand, to collect ever more information or facts, cyclically leads to a situation in which the new information directly undermines the existing foundation. Being simultaneously attracted and repelled by the desire for new experiences or information, these individuals have necessarily made choices as to what information would be taken in and what would be repelled.

From an evolutionary standpoint, these choices have created intellectual limitations because of the need for intellectual/emotional security. Given the attraction/repulsion dynamic, these individuals will experience cyclic intellectual implosions or cataclysms in which the logical structure of intellectual organization disintegrates. Emotional security is affected, but they have learned how to adapt because of the inherent mutable nature of the Third House. The symbol for mutability is an upwards moving spiral. The intellectual/emotional implosion reorganizes and the metamorphosis produces a new foun- dation, a more inclusive point of view.

The intensity of the intellectual/emotional implosion is directly proportionate to the degree to which the individual has taken in the information. If they have totally identified with the information or idea, then the cataclysm will be total; the foundation seemingly com- pletely removed. On the other hand, if the new idea is just being con- sidered, identified with or absorbed in a minimal way, then the intensity of the implosion will be less. These cyclic implosions are always caused when, through evolutionary pressure, the intellectual framework has become limited or stagnated to the extent of prevent- ing further growth or expansion. The degree of resistance relative to the need for expansion determines the degree of the cataclysm.

To determine the degree of personal identification in any Third House Pluto birthchart, check for other contributing or mitigating factors connected to Pluto. For example, Pluto in Leo in the Third House square Mars in Scorpio in the Sixth House would lead to heavy and intense identification and, therefore, resistance to chang-ing the intellectual organization and opinions. Pluto in Leo in the Third House trine Mercury in Sagittarius in the Seventh House would not create the same intense degree of personal identification, but rather a receptivity to new ideas.

The evolutionary pressure of the Third House Pluto to take in new information and to generate new experiences has created a rather intense degree of perpetual restlessness that resonates within the Souls of these individuals. This restlessness, linked to boredom and stagnation has led to the formation of an archive of experiences and information to which these individuals constantly refer, to increase their understanding and to communicate this understanding or knowledge to others. In other words, restlessness compulsively drives these people onward for ever new information or experiences.

Third House Pluto individuals have, and will need, relationships with others in order to process themselves; that is, to release the intellectual/emotional build up of mental energy. In extreme cases some of these people may become compulsive talkers. They need relationships not only to process themselves, but also to bring in new information from others. This continual interaction with others brings a constant need to adjust their ideas, which means that more and more facts are needed to explain in even greater detail the "com-plete picture" of themselves and the world in which they live. The problem is that these individuals feel that they will never know enough. Consequently, the "data banks" run the risk of becoming so large that they threaten to topple and disintegrate. Disintegration happens by having too many facts that logically connect in so many ways that there is no one composite or holistic way in which to relate all the data. Chasing one interest, idea, or desire after another, these individuals can end up walking down many different roads at the same time in a compulsive and unconscious effort to find that *one* fact or piece of information that will put it all together for them.

On this basis, these people can end up in a land of revolving perspectives with no center or foundation through which to create a consistent composite picture. The individual's inner experience of their own "center" is one of revolving perspectives, of a revolving mass of emotions, and the feeling that the center or foundation is always moving. Feeling lost and insecure, they can desperately reach out for one concrete fact to identify with and call their own. And yet, from an evolutionary point of view, the inner center of these individuals *must* move. The center or foundation is the movement itself. This spiraling effect can be endless as the perpetual cycles of contraction and expansion occur. The evolutionary basis for the constant movement of the center is to eliminate intellectual limitations that periodically form in the attempt to stabilize and feel secure. In addition, these cycles of contraction and expansion occur to progressively lead the individual into the awareness that there is an intrinsic limitation to what the empirically oriented mind can know in and of itself.

With Pluto's polarity point in the Ninth House or Sagittarius, the evolutionary intent for this life is to develop the intuitive faculty as opposed to an intellect that is primarily concerned with empirical facts. By developing the intuition, these individuals will progressively understand the deeper meaning or significance of the facts; the metaphysical or cosmological laws that are the basis of the physical laws themselves. They will learn how to synthesize all the data into a composite or holistic understanding of how the parts are related to the whole. Intuition is the faculty in all of us that becomes aware, or is aware, of knowledge that is not a product of deductive thinking. Intuition is that part in all of us that knows what it knows without knowing how it knows it. The necessary development of intuition for Third House Pluto people demands that they learn how to quiet their constantly busy minds. By doing this these individuals can open up and tune in to the intuitive faculty which is naturally plugged into the "truth" that exists in our universe.

A primary way in which these individuals can harness and quiet their minds is to align themselves with one comprehensive philosophical or metaphysical system that makes the most sense to them on a gut level. By doing so, these individuals can intuitively align themselves with one holistic principle upon which all the pieces of information floating around in their heads can be given a bottom line. By creating a bottom line all the facts can be synthesized in order to reflect the "truth" of the particular philosophy they have identified with. By aligning themselves with a comprehensive philosophy that is peronally identified with on a gut level, these individuals can create a consistent interpretation of the facts and the world around them. A quieting and harnessing of the mind will result.

Relative to the past evolutionary and karmic conditions, some individuals will create their own own bottom line philosophical or metaphysical principles upon which they base the arguments and "rightness" of their opinions and facts. These types will not align themselves with a philosophical system outside themselves. Others will align themselves with two, three, or more "systems" outside themselves. In so doing, they will select the principles in each that agree with some pre-existing point of view that they refuse to give up. In this way, they will synthesize these systems into their own brand of philosophy and truth. Others will make a total commitment to one philosophical system outside of themselves that they feel most drawn to in order to create the necessary bottom line. The voice of personal authority, with respect to their opinions, will thus be linked to the authority of the system itself. In all cases, the evolutionary need and pressure to seek out the bottom line cosmological or metaphysical principles upon which the facts are based is the same. This evolutionary pressure is teaching all of these people the connection between the physical and metaphysical, between the mind and intuition, and the limitations implied in the empirically oriented intellect.

Through the polarity point of the Ninth House, the individual must intuitively learn to realize the truth that explains the basis of what can be empirically observed. In other words, the archetype of the Ninth House correlates to the laws that are the very basis of Creation, the very basis of why human beings have developed religions, beliefs and philosophies. Just as belief systems can be limited or sectarian in nature, i.e. one version being "more right" than another version, so too do the Third House Pluto individuals run the risk of limiting themselves to a version of metaphysical truth by responding or reacting to the evolutionary requirements in the ways mentioned above. On the one hand, it is necessary to understand intuitively the bottom line metaphysical principles upon which the facts are based. On the other hand, to consider that those principles are the only ones that are relevant and "right" will promote a sectarian attitude. The narrow viewpoint is linked to the need for emotional security of an intellectual nature, and often implies a need to convince or convert others to one's own point of view. This implied limitation will thus guarantee, from an evolutionary point of view, that internal and external confrontation will continue to occur until the individual realizes that his or her version of the truth is just that: a version. It may be a necessary and relevant version given the individual's evolutionary and karmic background and requirements. The version is not the question in and of itself. Considering the version to be the *only* valid system is the problem. Once the individual is forced to make this realizaiton, or intuits the problem, then two things will happen: A. the individual will not need to defend personal opinions and philosophies to anyone, nor deny another's. Thus, the resulting metamorphosis will allow the individual to participate in conversations of all kinds in a nondefensive way, and to learn how to learn from others as well as others learning from him or her, and B. to continue the evolutionary journey in the discovery of the whole truth.

This journey will ultimately lead the individual into the awareness, intuitively realized, that the truth exists in and of itself. It always has, and always will. In this realization the individual will understand that the empirical mind and intellect are there to communicate and explain the truth. In other words, that the intuition applies to intellect and the intellect does not know what is true in and of itself.

In addition, the Ninth House polarity point will be teaching all these individuals the differences between reactions and responses. Prior to the necessary metamorphosis these individuals will be more or less a reactive type. In conversation, for example, they may not really listen to another, or they may wait for an idea or thought to react off. By not really listening, these individuals are concerned with impressing themselves on other people with their own ideas and knowledge. By waiting for a thought or idea to react to, these individuals are waiting to use another's thought or idea as a launching pad to assert their own point of view. Of course this dynamic can lead to conversations that go nowhere, or end up in inverted spirals of clashing perspectives. It can lead into arguments or intellectual confrontations in which all parties concerned try to convince or convert each other to their own points of view without seriously considering or listening to anyone else. In other situations, this reactive process can lead these individuals into simply walking away from a conversation wondering if he or she could have said something else, or said it in a better or different way. Or the individual could simply shut the person off and just walk away from a conversation. In all cases, the intent and effect is the same: to stimulate awareness of their own intellectual traps, games, limitations, motives and the needs and dynamics that have created them.

As this evolutionary drama unfolds, these individuals can learn the lesson of response versus reaction. A response is considered action that translates in this case to knowing when to talk and when not to, what to say versus rambling or trying to prove a point, and listening to other peoples' ideas and knowing what to take in and what to reject. Rather than buying a load of books that are not read, or are partially read, these people will learn to buy one book at a time within a certain subject area that they feel intuitively drawn to. Progressively, from an evolutionary point of view, these people will come to realize the difference between opinions and that which is true.

The four natural evolutionary conditions determine the kinds of ideas and philosophies that these individuals gravitate toward. Those in the herd state will listen to consensus opinions and ideas about any subject and would consider this point of view their own. They would gravitate to the "accepted" forms of religious or philosophical expression.

Those in the individuated state would desire to think for themselves. They would reject consensus opinion and would explore for themselves the subjects that attract them. They would tend to reject unquestioning attachment or bonding to any pre-existing formal philosophy, religion, or cosmological system. They would create their own system, or synthesize concepts from many different systems into a composite whole. Some individuals in this evolutionary state will select one pre-existing formal system that they feel intuitively drawn to. This is a sign that these people are moving rapidly toward the spiritual evolutionary state because they are beginning to realize the evolutionary and karmic limitations in creating and relying upon a metaphysical system of their own making. In other words, truth is to be realized, not created from an egocentric point of view. By aligning themselves with a system outside of themselves that reflects what they intuit to be true, these individuals are admitting that there are forces or powers larger than themselves.

Individuals in the spiritual state itself would attempt to understand the timeless basis for the physical laws of our world, and would use facts to illustrate a spiritual point or principle. These individuals will align themselves with specific cosmological, metaphysical, religious and philosophical systems that they feel most drawn to intuitively. These systems will be based on the intrinsic metaphysical truths and laws that are the basis of Creation itself. These intrinsic truths and laws are commonly communicated by these people, and the system itself, through metaphor, analogy, parables and inference wherein an empirical fact is used to illustrate the principle, truth, or law. Even within this evolutionary state, limitation can exist when either the specific system, or the individual, denies another system as "wrong". Evolution occurs in this state by progressively teaching the individual the whole 'truth', and the essential unity of all paths leading to the realization of the truth.

Once the necessary evolutionary lessons are developed, the Third House Pluto individual will have natural communicative skills that can inspire, motivate, enthrall, hypnotize, and metamorphose other peoples' intellectual patterns and opinions by sheer contact with them. Their ability to understand the connection between the macrocosm and the microcosm, between truth and fact, is unsurpassed. This ability can translate into taking any number of facts of a diverse nature and relating them to one central principle from which the facts emanate. The penetrating intellect can identify the essence of any question, problem, or issue that is posed by, or to, the individual. Once the core of any problem, question, or issue is exposed, the solution can soon follow.

Common characteristics of the Third House Pluto individual include: deep penetrating mind, intelligent, curious as long as they can control the direction of the curiosity, the ability to recognize the weakest possible link in any argument, ability to uncover the facts, appear to be very logical yet intellectually/emotionally defensive when other peoples' ideas threaten their own intellectual order, natural power emanating from the hands. For those involved in healing, they make excellent massage therapists because the electrical nervous system in the hands is highly charged and magnetic. Such individuals can resonate with the electrical field of another in such a way as to trigger their own neurological impulses that "tell" them what the client needs.

Famous people with Pluto in the Third House or Gemini:
        Benjamin Disraeli
        William Butler Yeats
        Queen Mary of England
        Napolean Bonaparte
        Robert F. Kennedy

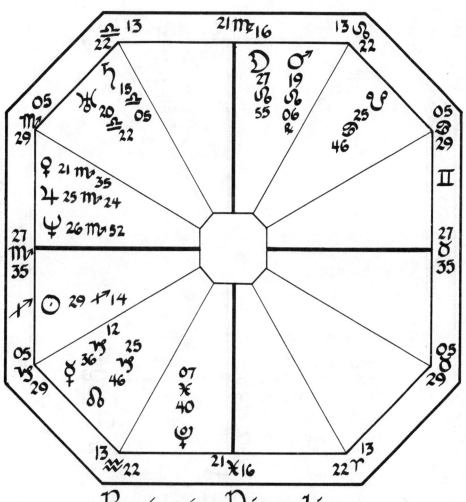

Benjamin Disraeli
source: Lois Rodden

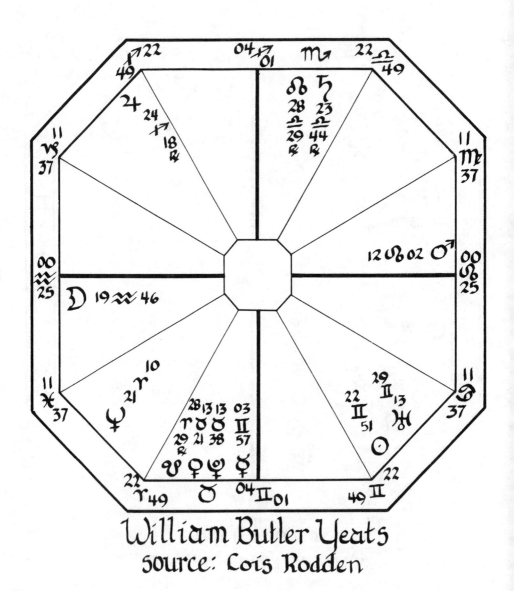

William Butler Yeats
source: Lois Rodden

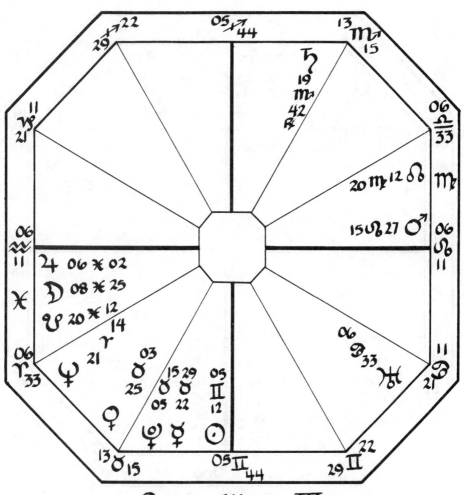

Queen Mary III
source: Marc Penfield

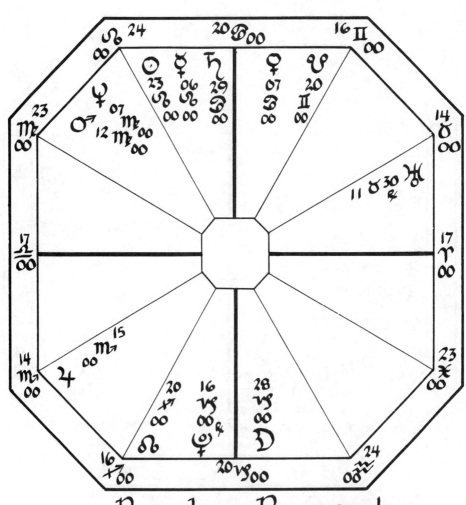

Napoleon Bonaparte
source: Marc Penfield

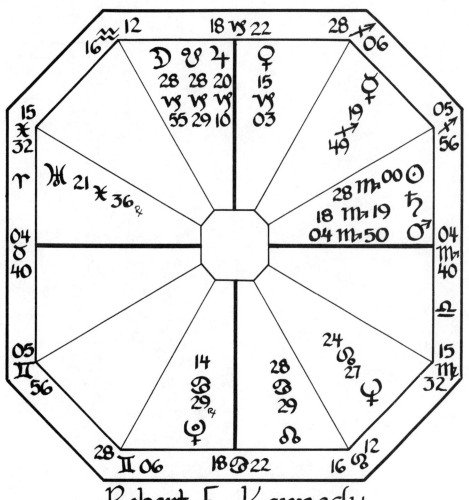

Robert F. Kennedy
source: Marc Penfield

PLUTO IN THE FOURTH HOUSE OR CANCER

Pluto in the Fourth House or Cancer demonstrates that an evolutionary cycle is closing and that a new cycle is under way. Individuals with Pluto in the Fourth House have been attempting and desiring to learn the evolutionary lesson of security rather than deriving emotional security from any external factor, such as parents, job, lover, etc., these individuals have been learning to create and supply their own emotional security from within themselves.

Because this evolutionary impulse is rather new, most individuals with a Fourth House Pluto have not completely learned this lesson. Thus most will choose to be born into family situations wherein one or both parents produce a variety of emotional shocks that force the individual back in upon him or herself. Commonly one or both parents of these individuals do not recognize, nor understand, the inherent individuality of these people, and therefore their emotional needs often go unfulfilled by the family. In this way these individuals are thrown back upon themselves to enforce the evolutionary lesson of internal security and the minimization of external dependencies. These lessons take place or manifest totally within the emotional structure of these individuals. The karmic and evolutionary lessons involve the nature and basis of the emotions, moods, and feelings — and the causes and sources of a particular mood, feeling or emotion.

This kind of early environmental experience can be extremely difficult for a child. As children, we are very sensitive and vulnerable because our sense of individuality, of who and what we are, is minimally defined. Our self-image, and therefore sense of individuality, is impacted upon in a very large way by our parents and early environment. We tend to take in wholesale the messages from our cultural surroundings, and from our parents.

About eighty percent of Fourth House Pluto individuals have had a series of prior-life experiences in which their emotional needs have not been successfully met by one or both of their parents. They will come into this life with those unconscious memories and, in many cases, will again be born into an environmental/parental situation in which their individuality and emotional needs are not recognized or met. It is not uncommon for these individuals to have a difficult karma or prior-life association with one or both parents in this life — unresolved karmic issues. These prior-life associations create memories that are held at an unconscious level. Thus, much of the interaction that can take place between child and parent is associated with compulsive emotional behavior patterns whose origins are rooted in or caused by difficult prior-life associations.

In cases of this kind, the specific nature of the prior-life difficulty or karma must be determined by relating the individual's birthchart to the chart of one or both of the parents. In varying degrees of magnitude, these individuals will have emotional problems directly linked to a feeling of not being sufficiently nurtured, appreciated, or understood for who they are as individuals. Coming into this life, the memory associations from the past are ignited once again as they experience the same kinds of environmental and parental situations. This karmic/evolutionary condition creates very intense emotional needs, as well as intense emotions, feelings, and moods. The unconscious expectation to have the parents meet these needs is the dynamic that creates the intense emotions, feelings, and moods when one or both of the parents fails to do so. It is important to understand that the individual's experience and interpretation of his or her parents' behavior can be quite different than how one or both of the parents interpret their own behavior toward the Fourth House Pluto person.

The inherent insecurity that many of these individuals arrive in this life with, creates a bottomless pit with respect to how much emotional nurturing that they require in order to feel secure. One or both of the parents may not be able to give the amount of emotional nurturing that the individual requires. The reasons may be very legitimate. For example, work or career concerns can remove the parents from the individual's immediate environment. As a child, the Fourth House Pluto person does not have the ability to understand the legitimate reasons of why one or both parents are not available. Thus, the individual interprets this situation as not receiving the necessary amount of emotional nurturing that he or she requires. This triggers the prior-life associations and memories, which in turn leads to the intense emotional states described above. However, from an evolutionary point of view, this is necessary because of the ongoing lessons in internal security and the minimization of external dependencies.

In other cases, some of these individuals will draw a parent who emotionally dominates their life. This is the type of parent who does not allow the individual to grow up. This type of parent desires to remain a central figure in the individual's life, and will covertly or overtly manipulate the individual. In this situation, the parent asserts his or her own emotional needs and does not recognize the emotional needs of the child. Again, the individual is thrown back in on him or herself. In still other cases, one or both parents will manifest cruel, vindictive, or outright abusive behavior toward the individual. This type of behavior can take many forms, and will manifest in varying degrees of intensity. We must determine the basis of the prior-life associations between the Fourth House Pluto individual and the parents in order to understand why these conditions exist, and how to resolve them.

In many of these cases, the intense emotional needs that have not been met in the early environment will translate into emotional displacement. In other words, these individuals can form relationships with others in which they unconsciously expect and attempt to have their emotional needs met by another. In effect, these emotional needs are those of a child. As a result, others can treat them as children and not allow them to grow up. All the emotional difficulties, and the pain associated with these difficulties, are compulsively projected upon others in varying degrees of intensity and expectation. The closer the Fourth House Pluto person becomes to another, the more intense the projections and expectations.

In intimate relationships, these individuals will unconsciously draw partners whose psychological make up is similar to one or both of the parents. This is especially true in the earlier part of their life. Through emotional displacement they are attempting to recover or experience, through perpetual demands, the early feeling of nurturing of their individuality that they did not have from one or both of the parents. Yet, because the partners are similar to one or both of the parents, this displaced need goes unfulfilled. Again, these individuals are forced back in upon themselves.

This cycle will repeat again and again until the Fourth House Pluto person progressively learns to understand why this kind of situation exists — learns how to supply their own internal security, and to emotionally mature through actualizing and establishing their own individuality through their own efforts. In addition, they must learn how to minimize or eliminate all external dependencies; of having their needs met by another, or having their need for security dependent on an external situation. Until this lesson is accomplished, the potential is great to manipulate others emotionally in order to have their childlike and self-centered needs met. Emotional tantrums (rage, anger), emotional withdrawal, crying wolf or feigning desperation in one form or another, extreme fluctuations of moods, deep feelings of guilt (What did I do to deserve this? Why did I just do that?) are common symptoms until they understand the lessons that they are learning, and the reasons for those lessons. As they learn to minimize their dependencies and resulting expectations, they will put their ongoing evolutionary lessons in motion.

In some cases, intense emotions and needs translate into an

utter fear of vulnerability. All Fourth House Pluto people have extremely sensitive and touchy emotions. In certain cases some of these individuals have learned how to cancel or deny their deep inner sense of emotional vulnerability and sensitivity. They have learned to do this through repeated lifetimes of emotional shocks, denial and emotional manipulation by their families and others. This conditioned reaction is also based on life experiences in which the famous rug has been ripped out from under their feet. The rug of emotional dependency will always be pulled out from beneath them when the degree of dependency creates a non-growth situation. As a result of this con-ditional emotional reaction, these people will experience emotional suffocation and frustration in this life because they are not expressing their emotional needs or expectations. Many of these individuals will even deny that they have strong emotions and needs; many will deny that they have any emotional problems at all. Yet, because the Fourth House person is intrinsically sensitive, vulnerable, and needing to be touched on an emotional level by at least one other, this conditional rection guarantees a lifetime (or cycles within a lifetime) of emotional suffocation, nonfulfillment, difficulties and a negative self-image because of the suppression involved.

Commonly, people who have reacted in this way will draw to themselves others whom they can dominate and control on an emotional basis. Typically the types that they draw will be emotionally weak or in need of some form of emotional healing. These others can themselves have unresolved or unfulfilled emotional situations revolving around one or both of their own parents. The Fourth House Pluto individual will control and dominate by identifying the weakest point in another's emotional makeup and will continually focus upon that weak point. In this way, they never allow the other to "get away" from their sphere of influence. They compulsively need to run the other person down, to tear them apart, in order to make the other feel as though the Fourth House Pluto individual is indis-pensable in their life. Thus, by putting another down, they are elevat-ing themselves, so to speak. And yet the other person is, in their own way, unconsciously placing these individuals in a position of dominance and control relative to their own karmic and emotional needs.

In this type of reaction the Fourth House Pluto individual guaran-tees the protection of their own vulnerability and essential insecurity on an emotional level. If this reaction is connected to unresolved anger or rage at one or both of the parents, the displaced emotional effect translates into taking this anger out on the other people that they draw into their life. This conditional emotional reaction leading to the need to dominate and control is a backhanded way of trying to fulfill their security and dependency needs. By being in a position of dominance and control they maintain their relationships and fulfill the security issue. Yet, their deepest emotional needs will go unresolved in this kind of conditional reaction because the people that they draw to them will be emotionally unequipped to help them. The point to consider is this: emotional denial does not equal emotional security, nor resolution of the evolutionary and karmic necessities.

This conditional reaction can also lead to situations in which these individuals attempt to dominate and control the environmental situations that they find themselves in. This kind of domination can occur through verbal/emotional dominance or through silently inject-ing a negative mood. Those who are manifesting a childlike response to their life, can project themselves in the same way. All too often this tactic will draw negative vibrations back to the individual, and will once again force them in upon themselves. On a positive note, this effect can lead to the awareness of the inner dynamics that are creat-ing these kinds of environmental effects in their lives. In this knowledge they can potentially make the necessary changes, and realize the evolutionary lessons involved.

The emotional reactions of the Fourth House Pluto individual can cyclically alternate between these two reactions. These individuals can alternate between seeking out places and people whom they unconsciously expect to fulfill their displaced emotional needs in a childlike way, or they can seek out environments and others whom they can dominate and control in order to have their unresolved emotional issues fulfilled in that way — to make others depend on them. These alternating cycles can occur within one primary relationship, or it can occur through the selection of different kinds of relationships or environmental situations at different times.

All Fourth House Pluto individuals will experience the full spec-trum of emotions, moods and feelings in the most intense possible way. Every emotion is deeply felt. Because the intrinsic nature of Pluto is to understand the why of anything, each mood/feeling/ emotion is intensely focused upon. This fixation can be so total that the person can appear paralyzed and unable to act in any way other than the particular emotion dictates. In extremely rare cases this dynamic will promote catatonic-like states. A classic example of this catatonic-like reaction is reflected in the last years of Fredrich Neitzsche's life, who had a Fourth House Pluto.

These moods/feelings/emotions originate from unconscious depths just like the seething lava beneath a volcano. Commonly some external experience acts as a trigger to induce the manifesta-tion of these phenomena. Sometimes the individual becomes a "victim" of some emotional wind that just seems to manifest out of the blue. Thus the individual can feel one way at a certain moment and then feel something else the next moment. These individuals can seem, as a result, highly inconsistent in their emotional expressions and needs.

Of course this dynamic can be just as confusing to the Fourth House Pluto individual as it is to those that they interact with. These intense inner states *must* be experienced because they are the source of self-knowledge and the path to inner security. The challenge is to become aware of the trigger or stimulus producing the emotions/mood/feeling, rather than just living it through without reflecting upon the origin or cause. If these individuals make the attempt to understand the cause or trigger behind any given emotional state, then they can develop a penetrating insight into the nature of emotional dynamics. This insight can be applied to themselves and to others.

These individuals will have two very distinct emotional cycles. On the one hand, they will have a cycle wherein they are deeply withdrawn, silent, and desiring to be left alone and undisturbed until they become aware of what is causing a particular feeling. They must withdraw because the origin of the emotion is manifesting from subconscious realms. Once the awareness comes, the other emotional cycle will be triggered. In this cycle they will become animated, processing whatever it was that manifested from the subconscious realm. These are natural cycles for these individuals. These alternating cycles can fluctuate wildly; they are inconsistent, unpredictable, and can change at any moment. It is important for these individuals to be involved with others who understand that they operate in this way. Extreme problems and emotional confrontations can occur when others in the individual's life do not understand this natural process. This lack of understanding by others can lead to emotional scenes and confrontations wherein they attack the Fourth House Pluto person in their necessary cycle of withdrawal. Others may read these cycles of withdrawal in an emotionally defensive way. They can feel threatened and insecure when these individuals withdraw and become silent. This will only aggravate the emotional reactions of the Fourth House Pluto person because others are interfering with this natural cycle.

A very common source of the emotional states and difficulties affecting internal security and self-image issues is that most of these individuals have switched gender in the most recent, if not the last, life. A new evolutionary cycle has began in which these individuals are experiencing, for the first time in a long time, the opposite gender than what they have just been. The male has been female, and the female has been male in the most recent prior lives.

The Fourth House, Cancer, and the Moon all correlate to the Jungian psychological dynamic called the anima/animus. The anima/animus dynamic, correlates to the basic fact that all people are intrinsically male and female. From an evolutionary point of view, each individual must experience the inherent polarities of maleness and femaleness. There is a natural and distinct law to the principle and experience of femininity and masculinity. Because there is a distinct law and principle to each, the psychological and experiential orientation to life will be naturally different relative to what gender an individual manifests through. Through natural evolutionary progression, over many, many lifetimes, all of us will learn how to unite both halves of ourselves in a totally integrated way, rather than living or playing out the extremes of each to the denial or detriment of the other. Pluto in the Fourth or Tenth House will correlate to this "gender switch". Other astrological symbols that can correlate to this gender switch are the South and North Nodes in the Fourth or Tenth Houses, or the nodes being in Cancer and Capricorn.

A feeling of security linked to gender can also be difficult to attain for the Fourth House Pluto person. Many of these individuals will not relate to themselves well in their present gender, especially considering the conditional ways that particular cultures define masculinity and femininity. Because most of these individuals have been relating to themselves in the opposite gender in the most recent prior lives, the hormonal and gender switch becomes its own source of emotional conflict. This conflict, again, is based on the fact that the emotional, intellectual, spiritual and physical structure of man and woman are intrinsically different. This conflict can be reduced in magnitude if this is not the first or second time the person has come back in that gender. This process is similar to buying a new pair of shoes. The more the shoes are worn, the most comfortable they become. They become our shoes. In certain cases this condition can be aggravated by one or both parents. As an example, a client of mine who had a Fourth House Pluto was brought up to be a female, even though he was male in this life. The mother wanted to have a little girl. Her anger and resentment of the fact that her child was male manifested in this type of behavior toward this man. Obviously, this early environmental impact devastated his self-image and emotional structure.

All of these factors contribute to the continuing evolutionary lesson of inner security. These individuals are learning to know and trust themselves so that they can know and trust others. Until they trust and know themselves, they cannot really know and trust another. In fact, many of these individuals will come into this life naturally mistrustful and suspicious of others and their motives. This lack of trust is based upon having been dumped on in situations or with people in whom they had become overly dependent. For many of these people, this lack of trust is also based on unconscious memories of experiencing emotional rejection or nonresponsiveness by one or both parents in the most recent prior lives. And many will have similar experiences again in this life. The only real antidote for this condition in this life is for these individuals to get to know and trust themselves, i.e. the nature and basis of their emotions and self-image, and to stop looking for security through external situations or other people.

Some individuals with the Fourth House Pluto will be exceptions to the above situations. Through prior-life efforts these individuals will have already learned the lessons of inner security, the minimization of external dependencies and the acceptance of their own unique self-image. Generally, about twenty percent of Fourth House Pluto individuals have made real efforts to learn these lessons before. Such individuals will be born into a family in which one or both parents contribute positively to these ongoing lessons. These will be parents who promote a positive self-image, encourage the individual to stand on their own two feet, and do not permit excessive dependency.

In some instances, the Fourth House Pluto individual will be born into a family that does not supply this support, yet the individual is not negatively impacted. These individuals will simply use that kind of family and environmental situation to further their evolutionary needs. Even in these cases, however, the emotions, mood and feelings described earlier will occur in varying degrees of intensity and frequency. In such individuals these emotions, moods, and feelings will occur for two reasons: 1. when the nature and structure of their reality becomes stagnated, and 2. when internal or external environmental circumstances manifest in such a way as to trigger an unconscious memory based on other lives. A case history will serve to illustrate this process and condition.

This individual has a Fourth House Pluto in Leo, South Node in Virgo in the Fifth House, and the ruler Mercury in Capricorn in the Ninth House. This individual was in the individuated/spiritual evolutionary condition. She was born into a family in which both parents could not recognize her individuality or emotional needs. The parents themselves were in the herd state evolutionary condition. She simply used this kind of family to continue her own evolutionary lessons.

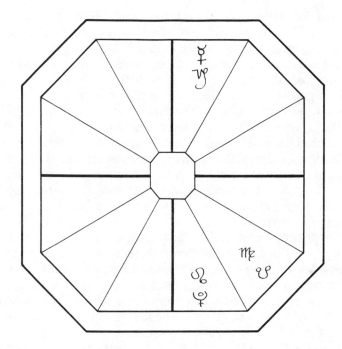

With the South Node in Virgo in the Fifth House, and Mercury in the Ninth House in Capricorn, she was naturally introspective, self-analytical and critical, and intrinsically oriented to understanding herself and life in metaphysical and cosmological terms. Her sense of personal identity and self-image were linked to metaphysical and cosmological perspectives and intuitive feelings. On balance, this individual was generally a happy-go-lucky type because of her ability to see the humor in most things — ruler of the South Node in the Ninth House. She more or less knew why she chose or came through the kind of parents she had because of her evolutionary condition.

However, her emotional needs as a child were not met. Her father, for example, was an alcoholic. Early in her life this factor made her go deeply within herself. Subconsciously she felt that she was being punished in some way — South Node in Virgo. Thus, she felt she deserved and was responsible for this kind of family situation. She held no malice toward her parents because of this early denial of her emotional needs. What she did not realize was the impact that this early denial had on her emotional nature because of the inherently suppressive pattern symbolized by the ruler of the South Node being in Capricorn relative to the Fourth House Pluto.

Only when her father died did all these supressed emotions, feelings and the resulting moods surface. She was in utter shock at the depth and intensity of these emotions. And yet, this event served not only to let out these buried emotions, but to also show her how they had subconsciously dictated the type of partner she had become involved with. This partner was unable to give to her what she needed on an emotional level. In addition, her partner was emotionally unable to understand who she was, and what she needed as an individual on an emotional basis. This realization promoted an attempt to redefine her relationship. Up until this point, she even denied she had intense feelings, moods and emotions. It took this kind of emotional shock, one of the ways that Pluto instigates our evolutionary necessities, to make her aware of all of these issues. As a result, a deeper self-knowledge occurred for her.

The polarity point is the Tenth House or Capricorn. The evolutionary intent is one of self-determination, learning how to accept responsibility for one's own actions leading to emotional maturity, learning how to walk on one's own two feet, and learning how to integrate or establish one's own personal authority or individuality in the context of society or culture. These lessons can, and must, occur through the individual's work or career.

When the Fourth House Pluto person acts upon these evolutionary intentions the "child" becomes an adult. They will learn that the responsibility for their life situation rests within themselves. By developing inner security and minimizing dependencies ("somebody make it happen for me") these individuals will learn how to trust themselves, know themselves, and ultimately know others. They will also learn how to be healthy adult children (we all have a child in us) who are able to be vulnerable and sensitive in a positive way.

By learning how to accept the responsibility for their own actions, these individuals will stop blaming others for their self-created "problems" or life conditions. In this way, they can even learn to understand the role that their parents played in creating the necessary evolutionary lessons. Perhaps they will learn to see their parents as just people who have various strengths and weaknesses; that they are just people who happen to be their parents. Those in an individuated and spiritual state can realize the karmic conditions and issues involved, and resolve to correct the situation in whatever ways necessary.

By making a self-determined effort to actualize the right work or career, they will create a counter-point vehicle through which the work environment allows for a natural reflection upon the sources, origins, and causes of the feelings. Through this process the individual gains personal knowledge and progressively develops control over their emotions, moods, and feelings. As the causes become known through reflection, these individuals will progressively be less inclined to be swept away, thus out of control, with their shifting emotional states.

The degree of emotional empathy that Fourth House Pluto individuals can feel for others will allow them to help others understand their own natures. They can encourage self-determination for others, and can help others accept the responsibility for their own actions. In addition, they can help others understand the nature of their own emotional blocks, and help or encourage them to minimize their own dependencies. In advanced evolutionary conditions, these individuals can achieve a total and equal integration of their masculine and feminine components — the anima/animus dynamic. This evolutionary metamorphosis will produce individuals who are able to identify their goals in life and, via self-determination, actualize them through their own efforts.

Common characteristics of the Fourth House Pluto include: emotionally intense and demanding (overtly or covertly expressed), cycles of depression and optimism, can be emotionally manipulative, insecure, defensive, easily threatened, potentially cruel, mean, or vindictive if deeply wounded, but also empathetic, sympathetic, nurturing in positive cycles, intensely loyal to those close to them, needing to control their personal environment or space.

Famous people with Pluto in the Fourth House or Cancer.
  Mia Farrow
  R.D. Laing
  Norman Mailer
  Fredrich Nietzsche
  Wolfgang Mozart

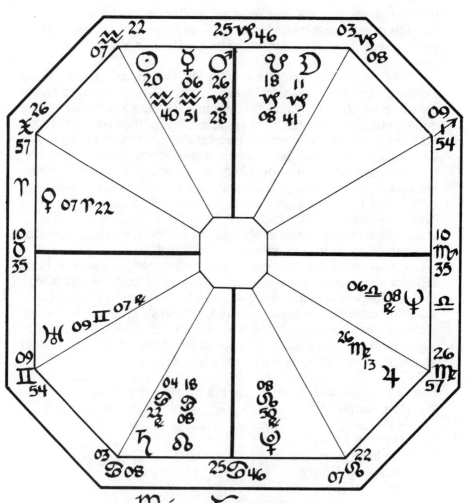

Mia Farrow
source: Lois Rodden

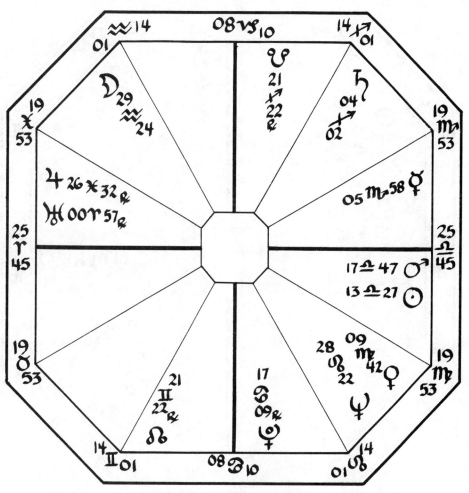

R. D. Laing
SOURCE: Lois Rodden

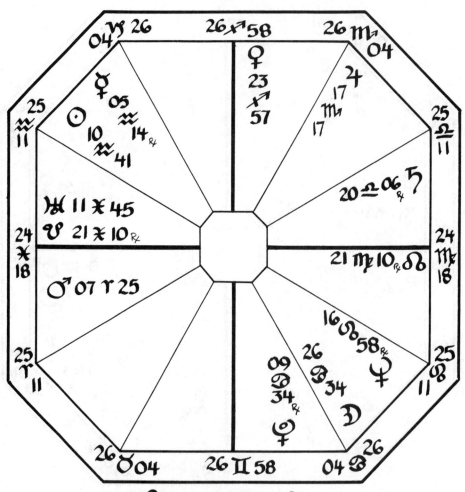

Norman Mailer
source: Lois Rodden

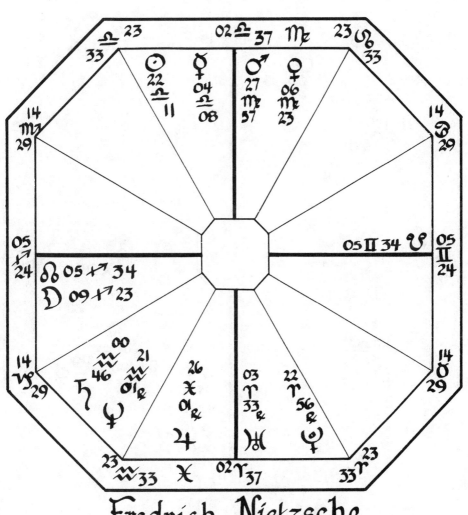

Fredrich Nietzsche
private source

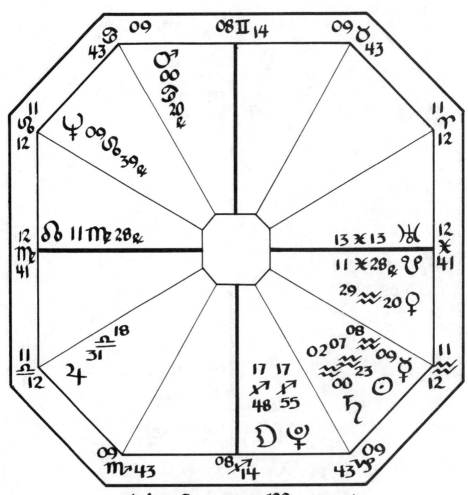

Wolfgang Mozart
source: Lois Rodden

PLUTO IN THE FIFTH HOUSE OR LEO

Individuals who have Pluto in the Fifth House or Leo have been learning the evolutionary lesson of creative self-actualization. The Fifth House and Leo are naturally fire oriented archetypes. As such, these people will feel as though they have a special destiny to fulfill just as those with Pluto in the Fifth House or Aries do.

Individuals with Pluto in the First House or Aries needed to maintain freedom and independence in order to discover their special-ness through experiences they felt instinctively drawn into. They were in a condition of perpetual discovery on a moment to moment basis. Individuals with Pluto in the Fifth House or Leo have needed, and will need, to actualize this special destiny. They will need to establish and project this creative purpose upon the environment. This creative purpose is no longer a matter of discovery, but is already formed in the Fifth House Pluto — it is inherent and complete. As a result, these individuals will deeply sense this special purpose in the depths of their Souls. The evolutionary desire and need has been, and will be, to fulfill this purpose from within. On this basis, these individuals have also been learning to take charge of their lives, and to seize destiny and shape it out of the strength of their wills. In other words, these people have needed to direct their own plays.

Because Fifth House Pluto individuals emerge from the womb with this deep feeling of specialness, they have desires and needs to be recognized and treated as special and unique. These feelings are so intense that they can create a virtually bottomless pit in terms of the need for love, attention and flattery. Whatever is received is never really enough. Although the needs can be temporarily satisfied, the compulsive emotional craving for attention, love and adulation drives the individual ever forward for more.

Children with Pluto in the Fifth House can create unique tests for their parents because they can go to any length to get recognition, love, and attention even if it is negative attention. These children can unconsciously manipulate situations, parents and others in order to receive the required attention. They can feel highly threatened or defensive when another child or person is given attention or love, or when the parents themselves show affection toward one another. Of course, this type of emotional behavior follows the person into adulthood.

People with Pluto in the Fifth House can be highly self-centered and self-focused. They can feel as though the universe revolves around them. Just as the Sun is the center of our solar system, individuals with Pluto in the Fifth House (the Sun naturally rules the Fifth House and Leo) can expect to have all things revolve around them. Yet these individuals must learn to understand that this condi-tion has been the evolutionary intention, desire and need from the past: to creatively actualize and be in charge of their own unique and special purpose. We must strive to understand the Fifth House Pluto individual in this light, and resist the temptation to harshly judge this kind of emotionl behavior in the earlier years of their lives.

Even though this orientation will be the theme in all Fifth House Pluto individuals, the variety of mitigating factors that condition and qualify the behavioral manifestation of this orientation will lead to many different kinds of expression. As an example, an individual who has Pluto in Leo in the Fifth House, the South Node in Gemini in the Second House, and its planetary ruler, Mercury, in Taurus also in the Second House will be self-contained, self-reliant, withdrawn, quiet, and undemanding. This individual has been learning how to identify his or her unique and special purpose or destiny by withdrawing into the self and learning to actualize the special destiny or purpose via their own efforts. This individual would know that they are special, and would not require an intense degree of emotional feedback or attention from others in order to have this feeling and knowledge validated from outside. In fact, the person would be intensely private, and would be able to fulfill most of his or her needs from within. With the South Node in Gemini in the Second House this individual would think about a variety of possible ways to best capture and actualize his or her creative essence. This process will revolve around inner considerations because the planetary ruler of the South Node is in Taurus in the Second House. Once the realization came as to the best possibility for actualizing the creative essence, this individual would begin to set the process in motion through his or her own efforts. This would be done in a relatively quiet way; a minimum of fanfare would be necessary.

We can contrast this example with Pluto in Leo in the Fifth House, South Node in Aries in the First House, and its planetary ruler, Mars, in Scorpio in the Eighth House. Given the feeling of a special destiny and purpose this person would compulsively manipu-late people and situations in order to be acknowledged and recognized as powerful and special. The degree of egocentric focus would be very intense. Beyond the need to be acknowledged by others, this individual would also need to prove and assert their power and specialness over others. In this condition, the individual would feel intensely threatened by the power and creativity of others, and would become defensive when others were given attention and ack-nowledgement for their accomplishments and needs. If the required attention, feedback and acknowledgements were not given, then this individual could inwardly react with anger and rage. As a child, this reaction could translate into intense emotional tantrums and demands. As an adult it could translate into cycles of withdrawn emotional seething countered by cycles of external emotional demands in an attempt to manipulate situations and people to get the required attention. In the first example the individual would think before they acted. In this example the individual would instinctively act. Reactions by others would prompt contemplation as to why he or she was drawing such reactions. Through this dynamic of action/reaction, the individual could learn to adjust their actions, and could realize, through trial and error, how best to actualize and discover their special destiny and purpose.

In these two examples the evolutionary theme and necessity are the same. Yet the mitigating factors have conditioned the application and ways to fulfill those evolutionary requirements. With Pluto in the Fifth House or Leo the focus has been, and will be, upon the creative principle. Creativity can be expressed in any dimension of life. It can be expressed through whatever unique and special capabilities the individual has. Because the evolutionary desire and intent from the past has been to actualize and establish the special creative purpose, these individuals have necessarily required an intense inner focus and determination to shape their destinies with the strength of their own wills. The mode of operation as symbolized by the South Node, and the vehicles to facilitate the mode of operation symbolized by the planet ruling the South Node, will show how this has been shaped in the past. All Fifth House Pluto individuals will have this need and desire regardless of the evolutionary condition or state. However, the evolutionary state will correlate to and qualify what is and is not possible for them to do with their creative needs. It will also correlate to the areas or external life conditions that it will operate through.

One of the karmic and evolutionary problems of this dynamic is that many of these people will consider themselves as the source of their own creativity because individualized (subjective) power is at maximum development or expression with Pluto in the Fifth House or Leo. Thus, many will feel that they, and they only, are the directors of their own plays. From a universal point of view, this orientation obviously implies a limitation. All the mitigating factors, and the observed evolutionary condition must be assessed to determine to what degree an individual has identified with the creative principle in this way. In the worst scenarios, some of these individuals will con-sider themselves to be miniature gods who are able to create and destroy reality at will. In the above example, with Pluto in the Fifth

House and in Leo, South Node in Aries in the First House, and Mars in Scorpio in the Eighth House, this scenario would be extremely likely.

Conversely, with Pluto in the Fifth House in Leo, South Node in Pisces in the Twelfth House, and Neptune in Libra in the Seventh House, the antithesis would be true. If the individual were in the individuated/spiritual evolutionary condition, he or she would be most likely to consider themselves as co-creators of their destiny, and sense, feel, or identify universal energy forces within their creative expression. Negatively, this same individual could feel almost powerless to actualize their special destiny and creativity even though he or she sensed, knew, and felt it deeply within. This reaction could occur because of the awareness and experience of others around them who seemed more creative, and more in control of their lives, than they themselves seemed to be — South Node in Pisces in the Twelfth House, ruler Neptune in Libra in the Seventh House. In all cases, whatever the mitigating factors and karmic conditions, these Fifth House Pluto people have been, and will be, learning to take charge of their lives in order to creatively actualize their own special purposes and destinies. The need to emotionally unite with the creative principle is the same in all conditions so that this evolutionary need can manifest. How this is done, and the reasons and ways that are karmically determined, will be different in all cases. Because the bottom line in all cases is the Fifth House Pluto, all will feel a sense of special destiny at a core level. All will feel the need to be recognized as special in some way. All will feel the need to express and actualize their creativity and will feel that they must be in charge of their lives in order to actualize their destinies. When the individual does not receive enough recognition according to his or her specific requirements, then a variety of emotional problems can result. When the individual is either overly identified with the creative principle from an egocentric point of view, or is experiencing a false sense of powerlessness to actualize the creative principle, then the individual will create or draw conditions that have the effect of counteracting the problem of false identification either way. Of course, varying shades of these two extremes can manifest. The specific pattern or arrangement relative to how the person has identified, applied, and responded to the creative self-actualization principle will determine the specific kinds of life conditions that are drawn or created to further the development of this evolutionary need in this life. As an example, the South Node in Aries, Mars in Scorpio described above would require conditions in which the individual experienced ego blows from the environment in order to pop the balloon of self-inflation.

Many of these individuals with Pluto in the Fifth House or in Leo have had prior lifetimes in which they were recognized as special in some way. As a result, many of these individuals will have uncons- cious memories of being treated as special by others. In this life these unconscious memories can lead into a situation wherein they simply expect to be treated and acknowledged in the same way again. They expect that others will cater to their needs, that doors will open before them, that the red carpet will unfurl as they emerge upon the scene of life. These memories can condition and dictate what these individuals think and feel they need in this life. Again, some will manipulate situations and people to receive the desired attention and recognition. This can be done in a variety of ways relative to the specific way that the individual has responded to this evolutionary impulse. In the case of the South Node in Pisces, Neptune in Libra in the Seventh House, wherein the negative reaction was a feeling of powerlessness, the individual could manipulate situations and people to get the required attention by feigning a sense of victimization. In this way the individual would attempt to draw others into his or her life who could help make life happen for them.

As a result of this prior evolutionary intent, and the associated memories, it is not uncommon for many of these individuals to create what I call the pyramid reality structure in which they place themselves at the very top. Every other factor in their reality structure revolves around and serves them; every other factor is subordinate or secondary to the fulfillment of their own purposes. Again, this can be done in a variety of ways as illustrated above. Because the need for recognition, acknowledgement, and attention can be so com- pulsive, many Fifth House Pluto individuals are not beyond taking emotional gambles or risks to get the needed attention. With respect to the pyramid reality structure, if the individual is not getting the needed attention or recognition from the point of view of their inner reality references, then many can threaten to topple their existing external reality structures in order to get the required attention. Thus, for example, if an individual is already married and is not receiving the required love and attention that he or she feels is their due, then some will involve themselves in a love affair to get this type of emotional feedback. Such emotional risk-taking can take many forms but the need for power and recognition will always be the underlying theme. In the case of an affair, the individual, of course, runs the risk of being discovered. If the individual is discovered then the possibility of having the existing reality (the marriage) destroyed would exist. This kind of emotional shock, and the different forms that this shock can assume will serve to create the necessary blow. This blow will enforce upon the individual the necessary realizations as to where he or she was coming from, and for what reasons.

With reference to the pyramid reality structure, many of these individuals will unconsciously manipulate others in order to have their emotional needs fulfilled. In the example where the individual could seek out an affair, he or she could cater to the desires of the potential lover in order to fulfill his or her self-centered desires and needs. As soon as the need is fulfilled, then the person will terminate the emotional risk situation. As an example, a woman who came to see me for astrological counseling was an unwitting victim of such a Fifth House Pluto individual. The man was a very prominent individual, married, and had children. Apparently he was feeling less than satisfied in his marriage. He "needed" a love affair. My client was a photographer. The Fifth House Pluto person was getting ready to publish a book —her own wishes linked to her career. He led her on by promising that she could produce the cover for the book, by catering to and manipulating her in this way, he had his own desires met. When the affair ended, he asked someone else to do the cover of the book. My client, of course, felt used.

Fifth House Pluto people can also be quite given to shaping and controlling the lives of those around them, including their children. This can occur because they have been learning how to identify their own special purposes and actualizing them through the strengths of their own wills. Thus, it is only natural that they will project this developed inner capacity upon others since their own Souls resonate to this vibration. Keep in mind that this process will work in all Fifth House Pluto individuals in some way. Even in the negative case illustrated earlier, i.e. the South Node in Pisces in the Twelfth House with the ruler Neptune in Libra in the Seventh House, the individual could attempt to shape and control the lives of others in this way: to suggest to others that they can have their own self-centered needs and desires meet in the very same way that the Fifth House Pluto individual does, i.e. feigning victimization. This need to shape and control the lives of others can be overtly or covertly stated, and it will manifest in varying degrees of compulsion and intensity. In the worse cases, the individual will subconsciously play God as he or she attempts to direct the lives of others according to what they feel or think another should or could do. In positive cases, the Fifth House Pluto person will encourage the individual development of those around them because they value this need in themselves. Yet, even in positive cases it can be very difficult for the individual to resist the desire to encourage this development according to what they think is right for the individual.

Fifth House Pluto people can be very giving and generous. And yet this giving and apparent generosity is commonly extended only when it suits some personal need. Because there is such a high degree of self-focus, the giving is commonly not related to the actual needs of another, but to what they think the other needs. This emotional complex, of course, is relative to all the other mitigating factors in terms of its degree of fixity and the different lenses — the position of the Nodes and so forth — through which it will be expressed. This dynamic can be difficult for many Fifth House Pluto individuals to understand because they are giving within their own reality references. However, that is exactly where the problem lies. In other words, the giving is occurring from within their own reality references, and not others. This orientation can set the stage for emotional scenes and confrontations when others manifest negative feedback in relation to what is being given by the Fifth House Pluto individual. These scenes, when they occur, are necessary because they can serve to undermine the pyramid reality structure of the Fifth House Pluto individual. In so doing, the individual will begin to experience the effect of the Fifth House polarity point: the Eleventh House.

The evolutionary intent described in the Eleventh House polarity or Aquarius is one of developing an objective rather than a subjective consciousness and focus. The individual must learn to link their special destiny and creative purpose to a socially useful or relevant function. In order for this evolutionary lesson to be realized, many of these individuals will be blocked or denied in fulfilling their unique purpose. The blocking force will be the social structure itself. This situation can lead to tremendous emotional frustration and anger. Frustration is heightened because society apparently is not acknowledging or recognizing the Fifth House Pluto individual as special. They sense within that they have a special destiny to fulfill, yet are blocked in a variety of ways from being able to realize it. Even if they can succeed in some way of actualizing some aspect of their creative purpose they may still experience the lack of enough recognition according to their own estimations of how much they should have. Accepting their lot in life can be very tough as a result. The royal red carpet seems to remain furled, the remembered acclaim but distant memories. By being relatively relegated to the sidelines, these individuals are forced to learn the lessons of objectivity. In addition they are learning how to become detached from the pyramid structure of their inner reality. As a result, these individuals may come to realize how to link their purpose to the needs of the whole or society. In so doing, they can contribute to the relevant needs of the whole and their unique and special gifts can be individuals who will learn that others are just as special as themselves. They can give to society or another in the ways that are needed as a result of this evolutionary intent. These individuals are learning to be a member of the play, not the director of it.

In the example that was used in the beginning of this section, i.e. Pluto in the Fifth House in Leo, South Node in Gemini in the Second House, and ruler Mercury in Taurus in the Second House also, the North Node would be in Sagittarius in the Eighth House. The ruler, Jupiter, is in Pisces in the Twelfth House. This individual realized early in life that music would be the best form and outlet for his creative expression and actualization of his purpose and destiny. As he grew and matured, he realized that operating in a vacuum would promote emotional and creative stagnation although he tried to do just that in the early part of his life. As the necessary emotional shocks occurred relative to self-imposed isolation, he realized that the most effective outlet for his purpose and creativity would be to link it to teaching. And the best form of teaching for him would be with children (North Node in Sagittarius in the Eighth House, the ruler in Pisces in the Twelfth House with respect to the Eleventh House polarity). The more he followed this path, the more his own creativity manifested itself and the more his need for acknowledgement was met.

As this process unfolds the evolutionary metamorphosis will produce people who can now recognize the individuality and needs of their children. This metamorphosis will allow them to guide their lives objectively, rather than willfully creating their children's identities out of their own self-centered images. It is not uncommon for Fifth House Pluto people to have children who are very strong-willed, self-centered, and who are resistant to the dictates of their parents. The resulting collision of wills produces the emotional confrontation necessary to induce the lesson of objectivity and detachment. The children can serve as mirrors who reflect the deepest inner dynamics of one or both of the parents. This mirror effect can be quite stark for one or both of the parents, especially if they are emotionally resistant to accepting, admitting, or seeing these similar dynamics in themselves. The evolutionary lessons of the Fifth House Pluto individual must also be passed on and taught to the child.

By learning the necessary lessons of objectivity and detachment, Fifth House Pluto individuals can also realize that they are not the source of creativity; they are a channel for the expression of the creative principle in the universe. This fundamental lesson is very important. Until it is fully realized, the creative flow can cyclically dry up or be blocked in order to induce the progressive realization of the true source of creativity. In this way, these individuals learn how to acknowledge creativity and specialness in others without feeling threatened.

Once these evolutionary lessons are put in motion with conscious intent, these individuals can create something new and unique in whatever field of endeavor they are destined to fulfill. That which they create can have uncommon depth and power of a transformative nature. Sometimes this creation may be ahead of its time — the dead artist syndrome. Yet, eventually that which was created will become accepted. These individuals can achieve a degree of fame or acclaim with respect to what they do apply themselves to. They can act as natural leaders and pioneers, and objectively understand how to guide the individual development and actualization of those that they touch. In some, the natural leadership abilities is a sight to behold. The other day I was watching several children play. They were all four-and five-years-olds. At one point an eighteen month old baby wandered by these children. This little girl had a tremendously powerful aura. All the four and five-year-old children stopped what they were doing and began to follow her. Mesmerized they simply fell into line behind this child. The baby was undaunted, and simply kept walking. After watching this scene, I asked the mother of the baby if she would give me the birth data for her child. She humored me and obliged. As it turned out, the little one had Pluto conjunct Saturn in the Fifth House, Pluto being in the last degree of Libra conjunct Saturn in the early degrees of Scorpio. The South Node was in Capricorn in the Eighth House, the South Node's planetary ruler being Saturn. This intrinsic leadership capacity created an almost Pied Piper — like effect upon the four-and-five-year-olds even though she was only eighteen months old!

Common characteristics of Pluto in the Fifth House or Leo include: tremendous strength of will, dignified, creative, needing attention and acknowledgment, loving, giving, can issue attention and love upon others, magnetic, generally narcissistic, king or queen complex, powerful and intense at a core level, can be quite demanding in overt or covert ways, suspicious of yet needing flattery, very protective of those close to them.

Famous people with Pluto in the Fifth House or in Leo:
  Richard E. Byrd
  Sir William Crookes
  Sir Richard Burton
  Mao Tse-Tung
  Herman Hesse

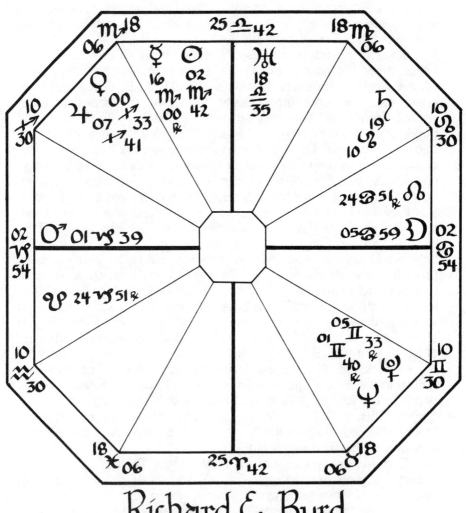

Richard E. Byrd
Source: Lois Rodden

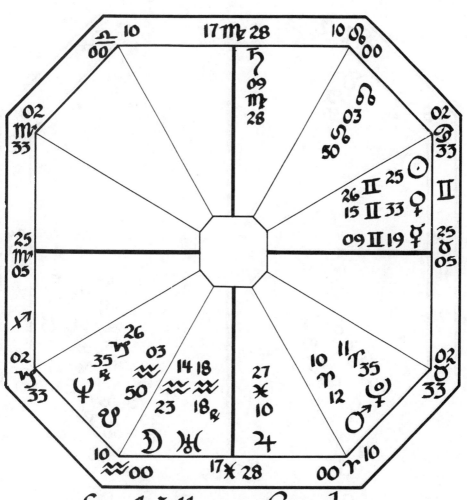

Sir William Crookes
source: Lois Rodden

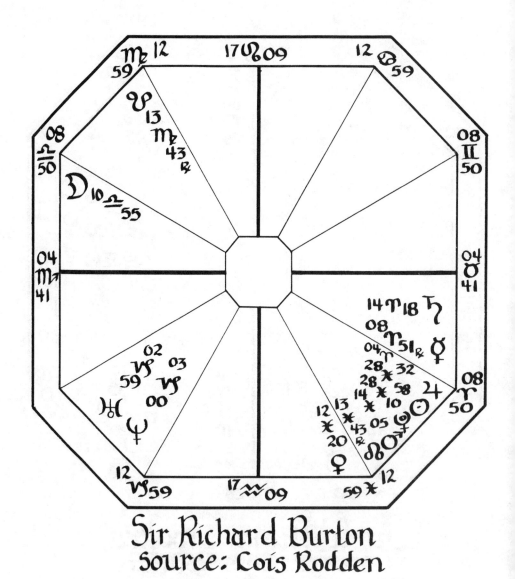

Sir Richard Burton
Source: Lois Rodden

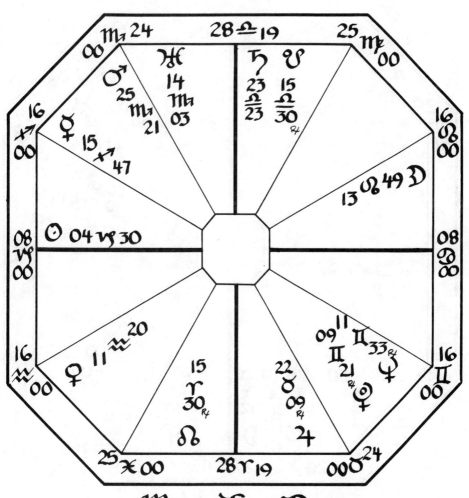

Mao Tse-Tung
source: Marc Penfield

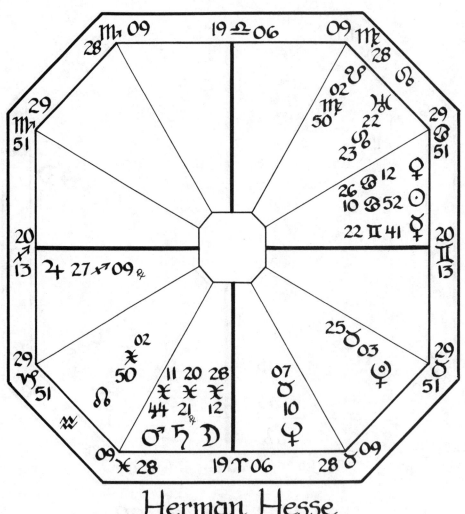

Herman Hesse
source: Lois Rodden

PLUTO IN THE SIXTH HOUSE OR VIRGO

Individuals who have Pluto in the Sixth House or in Virgo have been learning the evolutionary lessons of service to society, to culture or, on a smaller scale, to an individual. In addition, these individuals have been learning necessary lessons in personal humility and self-doubt, discrimination, purification and self-improvement. These specific lessons have and will be developed through intense self-analysis.

In the broadest possible sense, the Sixth House or Virgo is an archetype symbolizing how individuals within a collective body, i.e. culture or society, learn to fulfill a specific function within the culture or society. Each person, in any group of people from the most primitive to the most sophisticated, has a work-related function to assume in order to help sustain the entire group. On this basis, the individual's personal desires, needs, and times are "sacrificed" in order to fulfill the desires, needs and requirements of the tribe, culture, or society. This archetype teaches all of us the emotional, intellectual, spiritual, and physical dynamic of service to the whole. In certain cultural tribal, or societal contexts, the individual's work assignment or role is fixed according to hereditary status. In other cultures, people are free to determine their own work. In either case, the lesson of personal sacrifice and service is learned.

Those with Pluto in the Fifth House or in Leo learned how to identify the creative principle and purpose from within themselves. In order to do so, they necessarily had to unite emotionally with the creative principle in order to actualize their own purpose. This led to the creation of the "pyramid" reality effect in which the individual was on the very top of the pyramid, expecting all other factors in his or her life to revolve around and serve their needs.

In the Sixth House or in Virgo, the pyramid now becomes inverted. The individual is at the bottom in order to serve the needs of the whole through the work or service function that he or she has been assigned or has chosen. In either situation, the individual with the Sixth House Pluto has necessarily had to learn practical methods, techniques, or skills in order to establish or apply their work-related function within the culture tribe or society.

The Sixth House or Virgo is what I call a transitional archetype in which the individual is leaving behind the Leonian Fifth House orientation, and moving toward the Seventh House or Libran orientation: the archetype or lesson of individual equality or relativity. Aries through Leo, the First House through the Fifth House, is necessarily subjective or egocentrically oriented. Libra through Pisces, Seventh House through Twelfth House, is necessarily objective or progressively egoconcentrically oriented (the individual relating their identity to larger and larger wholes; to the universal).

In effect, then, the inverted pyramid effect has been teaching these individuals to pierce the balloon of their own self-inflated specialness or importance. The piercing of this balloon has induced necessary evolutionary lessons of humility and self-purification. They have been learning that the universe does not revolve round them, and that there is much more concerning the totality of reality than just the reality that can be seen from the center of their own pyramids. In the Fifth House or in Leo the individual was learning to actualize that which they were. In the Sixth House or in Virgo the individual has been learning to understand that which they are not. Being aware of what they are not, these individuals have not only been learning essential humility, but also a necessary self-purification that purges all traces of self-glorification and delusions of grandeur.

Again, one of the ways that this lesson has been learned is through the work dynamic. This has been learned in a variety of ways depending on the specific karmic signature and natural evolutionary condition of each individual who has Pluto in the Sixth House or in Virgo. As an example, let's put Pluto in Gemini in the Sixth House, South Node in Virgo in the Ninth House, and its planetary ruler, Mercury, in Aries in the Third House. Let's also say that this individual is in the individuated evolutionary condition. For the sake of our example, let's say that this individual decided to be a teacher in order to fulfill his or her work requirements and service to the whole. In this evolutionary condition it would not be uncommon for this individual to intuit and analyze the deficiencies in the ways, methods, and techniques of teaching that were learned during his or her own training. Thus, this individual would think about new and different ways of teaching in order to improve and advance the function of teaching in a generalized sense. The individual would be given to devising new techniques and methods of teaching: of what to teach, and how to teach it. With Merury in Aries, these new ideas could certainly pioneer and break new ground for the profession of teaching in general. The problem that would exist is one wherein the individual attempted to convince and convert others to his or her point of view. These others would be fellow teachers specifically, and the pre-existing organizational structure of the teaching profession in general. These new ideas, methods, and techniques would be considered as important, relevant, and necessary by the individual. The individual would be so personally committed to these ideas that they would represent his or her emotional security and personal power. Thus, the individual would need to convince and convert others in order to establish power and personal identity in the work place.

By concentrating and focusing on the pre-existing methods of teaching in a critical way, the individual would naturally draw criticism about these new ideas on how to adjust and improve the discipline of teaching. The new ideas, techniques, and methods would be too "new" from the point of view of the others because they would directly undermine and threaten their own emotional security, power and personal identity which are based on the existing ways that they have been teaching. As this individual emerges on the scene with new ideas, the necessary critical feedback that the person would receive would promote ongoing lessons in humility and self-purification. The new ideas would not be received with the positive acclaim that was expected, nor would the individual be perceived to be the 'savior' or 'hero' that he or she desired to be. The necessary negative feedback would thus serve to create self-analysis leading to the adjustment and revision of the new ideas, techniques, and methods. This process would require the individual to work within the ranks of the profession: not outside of it by hurling bombs of arrogant criticism. By working within the ranks, the individual would not only learn necessary humility, but would also learn how to adjust the very language used in order to produce the positive changes that he or she had the intrinsic capacity to create. In so doing, the necessary adjustments of humility, service to the whole via the inverted pyramid effect, and the purging of egocentric patterns of individual grandeur would be accomplished.

In addition to learning this lesson through the work dynamic and service, these individuals have also been learning it through mental self-analysis. Those with Pluto in the Sixth House or Virgo have a very powerful mental x-ray or laser-like mind. The intensity of the mental light exposes any dynamic, component, or part of themselves that is not "right" or "perfect." These dynamics, components, or parts can be of a mental, emotional, intellectual, or spiritual nature. The Sixth House and Virgo are intrinsically yin or feminine by nature. As such, this produces cycles of natural introspection and self-analysis. These cycles allow for a periodic mental and emotional survey of the individual. This survey must occur so that the individual can cyclically eliminate and adjust any component part within his or herself that is not attuned to his or her ongoing evolutionary and karmic needs. Again, the Sixth House and Virgo are transitional archetypes. Thus, that which must be adjusted or eliminated will pertain to internal patterns of egocentric self-identification, the pyramid reality structure, delusions of grandeur, or any intellectual and emotional pattern that is preventing further growth.

The intense cycles of self-examination that focus upon these internal patterns will correlate with inner feelings of what must be changed in some way. This inner experience clearly reflects a standard of conduct that is "right" and "perfect." Because of the transitional nature of this archetype, these individuals are evolving toward an egoconcentric orientation to life — to identify themselves with larger and larger wholes. That which is not right or perfect will be subconsciously linked to the life orientations symbolized by Aries and the First House through Leo and the Fifth House. That which is right and perfect will be symbolized by Libra and the Seventh House through Pisces and the Twelfth House. The vehicle through which this transition takes place is Virgo and the Sixth House. Thus the cycles of intense self-examination and analysis occur so that the necessary adjustments can be made. In so doing, the transition can take place.

Many of these individuals have memories stored at an unconscious level that pertain to prior indiscretions and mistakes, or anything that was not done "right." Again, we can see that a standard of right conduct promotes this inner feeling. In some individuals this standard of right conduct is highly defined at a conscious level. In others, the standard will only be sensed as a haunting feeling of what should or ought to be. Whatever the case is, most Sixth House Pluto individuals will experience guilt at an unconscious or subliminal level. This guilt is based on prior mistakes or indiscretions as measured against the standard of right conduct — of what should have been, or what should be. The basis of this guilt will be different in all cases, and will be described by the main karmic/evolutionary dynamic in each chart. In all cases there will be a desire and need to atone for the guilt in conjunction with the standard of right conduct. To atone is to purify and effect self-improvement. A way to do this is to serve others in some way — the work dynamic.

Because of the need to atone for guilt, these individuals will mercilessly analyze their own shortcomings, flaws or inadequacies, and truly become their own worst enemies. By compulsively focusing on imperfections, these individuals can become hung up or blocked from being able to actualize or fulfill their deepest personal capacities or abilities in this life. Somehow, they are never ready (perfect) or good enough. On another level, they can feel that they do not deserve more than they have because of the guilt. Thus, they can drag their feet, procrastinate, or create rationalizations (excuses) as to why they can't do something that they know they should, could, or want to do. By procrastinating and not fulfilling their full potential, they create yet another source of guilt. Many of these individuals have been carrying the burden of such guilt for many lifetimes: the knowing at some level that they have copped out on themselves in some way. This kind of evolutionary and karmic pattern can become a vicious cycle that perpetually feeds itself. In the worst cases, the individual will be in a state of perpetual crisis because of this evolutionary and karmic condition. The need to atone for the guilt that promotes the com-pulsive analysis of the individual's inadequacies, flaws, imperfections, and lacks can translate into the inner experience of "what did I do wrong?". Some individuals who externalize and project this inner orientation upon the environment will blame others for the "wrongs" that have been projected on them. They can feel victimized by the environment, others, and the nature of their circumstances. This kind of internal and external orientation is relative, and is experienced differently in each Sixth House Pluto individual. The difference is based upon the specific karmic/evolutionary signature in the birth chart.

In the example that was used earlier, the individual could have refused to make the necessary adjustments that would have allowed for the new ideas, techniques, and methods of teaching to be accepted by others. The negative feedback and criticism from others would be used by the individual to promote the inner analysis of what he or she was doing wrong. Yet, it would be simultaneously projected upon the others: the wrong that they were doing to the individual. Thus, the individual would feel victimized by the teaching environment and his or her co-teachers. The person could then feel guilty for not being able to actualize and establish the new ideas, techniques, and methods of teaching because he or she could or should have done so — a standard of right conduct. This experience would then enforce the analysis of what was inadequate, imperfect, flawed, or lacking within the individual that created this situation.

In effect, the need to atone for whatever the origins and causes of the guilt are can lead many of these individuals into a reality structure that is full of personal crises. Crisis is a necessary experience for these individuals. At one level, they may unconsciously hold themselves back from fulfilling their inherent capacities, and may only minimally develop or fulfill their abilities. When this reaction occurs, it will reflect the need for self-denial, self-penance, self-sacrifice, or self-doubt. This behavorial reaction creates the necessary crisis because the person will in some way recognize the above patterns and failure to fulfill the totality of his or her capacities. The ensuing emotional/mental crisis will cyclically enforce the need for self-analysis in order to expose the inner reasons or dynamics that have created this behavioral reaction.

Crisis brings these issues to a head. Positively responded to, the analysis of the issues can promote self-adjustment, self-improvement, and self-knowledge because the light of personal awareness, experienced through the crisis, allows for behavioral change. Behavorial change can occur once the inner dynamics and origins of the behavior are brought into the full light of day. Self-adjustment and self-improvement can occur by aligning with a standard of conduct that is right for the person, and by consistently practicing the methods, techniques, and ideas of that standard. In this context, the standard of right conduct could be any system of thought that promoted self-improvement and self-knowledge.

Negatively responded to, the analysis brought through crisis will promote a compulsive need for continual self-criticism to the extent that the individual is unable to implement constructive change: he or she simply gets hung up in the unending web of mental analysis which can create a kind of emotional and mental paralysis. Ancient Taoist philosophy offers a simple yet instructive example to illustrate and counteract this negative effect: The centipede has a thousand legs and is just fine as long as it keeps putting one leg in front of the other; it gets to where it wants to go. Yet, when the centipede tries to analyze how leg forty-six, for example, works in the sequence of one thousand legs, it freezes and becomes paralyzed. The antidote is to just keep putting one leg in front of the other.

In the same way, Sixth House Pluto individuals need to keep walking, talking and acting. Perfection or atonement will occur in this way. It will not occur by sitting in one spot and trying to become perfect, clean, and guilt-free from that one spot of inaction. The individual can not improve simply by thinking about it. There must be the initiation of actions that allow for the implementation of the motion so that a processing can take place. Even though the Sixth House and Virgo are intrinsically feminine or yin in nature, they are also mutable. Movement or action promote clarity and understanding of that which is occurring through inner absorption or analysis. The intrinsic motion of an action allows for a processing with respect to that which is being analyzed. The result is perspective and clarity of thought. Without the necessary motion implied in an action, that which is being analyzed can become distorted leading to a loss of perspective.

Sixth House Pluto individuals can induce another form of crisis; they can create so many external obligations that they have no time for themselves. Again, we see the theme of work and service linked to the inverted pyramid effect. We see this phenomenon linked with unconscious guilt/atonement patterns that are in turn linked to the need for egocentric purification and humility. Such behavior is also associated with personal avoidance. All too commonly these people feel absolute aloneness at an inner level. This acutely felt experience is a reflection of the mental/emotional analysis through which the individual is stripping him- or herself to the bone. This experience is also a reflection of the fact that the Sixth House and Virgo are a tran-sitional archetype. Progressively the individual cannot relate to nor identify with the Aries through Leo orientation to life. The individual is moving toward the Libra through Pisces orientation to life. This transition makes the individual feel alone because he or she can neither relate to the past, that which is known, nor relate to the future because it is unknown. Thus, the experience of the moment, the per-petual transition, leaves the individual feeling utterly alone at a core level.

By stripping him or herself bare, by analyzing all the component parts that must be adjusted and improved in order to be reorganized into a more perfect whole, the individual can project this analysis upon the external environment. In doing so, the individual experiences the imperfections, the inadequacies and the flaws of life in general. Such a critical analysis of the inner and outer environment produces the feeling that something is "missing", or that life is ultimately meaningless. This intense emotional focus, negatively expressed, translates into a few other common behavioral manifestations. In addition to being generally self-critical, the person may be critical of everyone or everything in their environment. This criticism, based upon some standard of conduct established by the individual, can translate into personal avoidance becuse the intense feelings of aloneness are just too great. Thus, the individual becomes very busy, filling up his or her life with one obligation after another, and never getting around to doing what they ought to do for themselves. The crunch of time, in relation to obligations or projects, enforces crises making the individual aware of the inner dynamics that create this behavior. Crisis can promote change if the individual so desires. In this context, it would mean a change of personal routines which would allow the individual to get in touch with him or herself at a core level. Then the necessary activities can be developed through which personal improvement can occur. In many people this dynamic of personal avoidance, when linked to the gut sense of an inner void, translates into the workaholic syndrome.

Other common self-avoidance habits are: always reading (nor-mally "trashy" or meaningless books), eating, or going out all the time. Those who continually complain and criticize cannot help but attract attacks or criticisms upon themselves. This, of course, will only deepen their sense of isolation. It can also deepen the sense of bitterness and meaninglessness. Such people can simultaneously feel sorry for themselves while vindictively lashing out at the misidentified sources of their problems. They may feel persecuted, or that "forces" are conspiring to attack them. They may in fact *be* attacked in some way because of factors arising from their own actions in this and other lifetimes. A few of these individuals with Pluto in the Sixth House or in Virgo are being forcefully held down by the weight of their own actions in the past; karmically crushed by their own actions in other times. This condition normlly occurs through some kind of prior abuse of power. Being forcefully held down enforces humility, as well as the lesson of hearing how to receive power versus desiring power for its own sake.

The key to resolving this crisis is to become aware of the reasons or dynamics that are creating this kind of experience. The reasons, lessons, or dynamics concern the evolutionary necessities of the Sixth House Pluto. Remember that these individuals will unconsciously "set-up" or attract situations in which they must be of service to the whole or others, or necessarily experience personal sacrifice or denial because of the inverted pyramid affect. Also keep in mind that the experience of all the dynamics that have been discussed so far is relative in terms of the intensity and karmic need. Each case is different but the evolutionary theme will be the same in all cases.

As examples to show how relative the experiences of these dynamics can be, we can put Pluto in Cancer in the Sixth House, the South Node in Pisces in the Second House, and its ruler, Neptune, in Leo in the Seventh House. In the spiritual or universal evolutionary condition, the individual would have identified with a spiritual system of values and ideas that served to promote self-reliance. In addition, they would use the values as guidelines to effect self-improvement with respect to that which is imperfect within them. This individual would realize that the source of their problems is totally within him or herself. In addition, the individual would already have learned how to give to others via the work that he or she was destined to do, and would use the spiritual ideas and values as a guide in doing so. This individual would embody a natural humility and use crisis as an opportunity to grow, realizing that external reality is but a metaphor for internal reality. In a very evolved spiritual state, this individual would be able to take on the karma of other individuals as a form of service and personal sacrifice to them, perhaps even taking on the physical ailments of others.

This same pattern linked to the individuated state would produce an individual who felt essentially alone coming into this life.

The individual could feel a high degree of alienation from "common" people and would have a hard time relating to them. The resulting crisis could revolve around not knowing what to value or think. The individual could attempt to solicit opinions, ideas, and values from a variety of people and sources, and could even try a variety of lifestyles and roles in an attempt to discover the meaning that he or she was seeking. Each new attempt to identify with this value or that, this idea or that, would produce a series of personal crises as the forms and limitations of each became nonproductive and meaningless. In conjunction with this, the individual could also try on a variety of roles or situations in an attempt to find the one that felt right; one that he or she could believe in. Again, as the roles or situations dissolved into meaninglessness, a personal crisis could ensue. The individual would criticize all that was "normal" as defined by society, and attempt to separate from the mainstream. In turn, the individual could receive criticism from society for being different.

In the herd state, the individual would attempt to live the values and beliefs that their cultural context was founded upon: that which most people valued and believed. The mundane and common forms of work would be vicariously associated with and lived out. The emphasis would be on that which was "practical" from a societal point of view. Any ideas, beliefs, values, or lifestyles that differed from consensus opinion would be criticized in varying degrees. In turn, the individual would attract criticism from those in the individuated state for being blind to other points of view, alternative values, and different ways of relating to themselves and others. The attraction of external criticism would mirror the fact that this individual was in some way feeling a lack from within with respect to their own values, beliefs, and lifestyle. The internal lack would reflect an inner crisis that made the individual aware at some level of something that was missing in life. Yet to try other approaches could also lead to a crisis of the unknown and the uncharted. Locked within the prison of their own complacency, the individual could become alienated from the very values and beliefs that he or she has been vicariously living out. Yet, fearing the unknown, the individual is unable to change, and can only criticize those who do not conform to that which is "normal" and "practical."

Another form of crisis is physical illness arising from self-denial, repression, emotional retention, blocking activity, or overextension of personal energy in some way. Physical illness may occur in order to induce the necessary confrontations that allow for an awareness of personal behavioral patterns, and the reasons or dynamics that motivate them.

The polarity point is the Twelfth House or Pisces. The evolutionary intent is one of simplification, of dissolving old mental and emotional barriers that are preventing a holistic and clear understanding of the individual's own self-concept. The polarity point promotes clarification of the relationship between the individual and others, an awareness of proper priorities. It also promotes the development of faith and the sense that the individual is connected to much larger forces than just him or herself. The polarity point fosters an understanding as to why the individual feels so essentially alone, and finally induces a transference from deductive to inductive reasoning.

The experience of essential aloneness enforces the lesson that nothing will fill the inner void other than a relationship to the Divine. Yet, many will deny any "cosmic connection", or ideas to this end: it's not "practical", or it is "irrational" is the common claim. However, the function of aloneness is to enforce this lesson at some point in the individual's evolutionary journey. Many of these individuals have approached spirituality in prior lives, but this approach has normally been associated with cycles when a crisis was so extreme nothing else would work except a spiritual perspective. Yet, as soon as the crisis blew over, their commitment to the spiritual point of view faded away with it. By attempting to learn this lesson, these individuals are also learning the difference between deductive and inductive logic. Inductive logic focuses upon the whole so that the parts reveal themselves. Deductive logic attempts to build the whole out of all the component parts. By learning to develop an awareness of the whole, these individuals will be able to understand and experience themselves in a more complete way, rather than isolating a particular feature and attempting to understand themselves through that feature alone.

By cultivating a more holistic view of themselves and life in general, these individuals will mentally and emotionally be able to simplify themselves and their understanding of life. In addition, they will develop a clearer vision of their relationship to other people, of when to serve and when not to be of service. Thus, an essential lesson of discrimination is learned.

Cultivating a relationship to the Divine brings lessons of self-forgiveness, as well as tolerance and respect for the mistakes and imperfections in others. Cultivating an awareness of the whole develops an inner peace that will replace compulsive anxiety and worry. An awareness of the whole enables such individuals to see or sense what their correct or proper work function is within the whole, rather than doing any job just to work. Proper work will always be found in relationship to natural abilities and tendencies, which derive from prior-life efforts.

By actualizing the proper work function, these individuals can further expand the horizons of personal awareness. In spiritual terminology, we have the idea of karma yoga. By cultivating an awareness of the Divine, or the whole, these individuals will purge or purify the ego in such a way as to allow the universal spirit or current to operate through them. Sincere and real humility will be manifested, rather than false humility, which is no humility at all. By learning these lessons, these individuals will flow with the natural currents of life, rather than analyzing why the currents go this way and that. The centipede will walk on.

Those who totally resist this evolutionary intent and maintain the old mental and emotional patterns because of a compulsive need for crisis and a sense of victimization can truly lead lives of pain, misery, sickness, and the sense of being in a living prison of their own mental chaos. Those who totally move with the evolutionary intent will surrender themselves to the universal spirit. Allowing the spirit to use and direct them, they will possess an inner luminosity that transcends "rational" explanation. The essence of humility, such people will be of service to anyone who is in sincere and legitimate need of their help.

Those who move from one side of the line to the other, from resistance to cooperation with the evolutionary intent, will experience cyclic moments of extreme clarity, and moments of great confusion, discouragement and crisis. The intensity of each are dependent on the amount of resistance and cooperation to the evolutionary intent. Within the extremes, many of these people will simply find their meaning in life as reflections of the particular "duties" or functions that characterize their lives.

Common characteristics of the Pluto in the Sixth House or in Virgo include: keen analytical mind, self-effacing, willing to help others, naturally shy or timid, extreme sense of self-consciousness, critical or forgiving of others depending on circumstances, always busy with one thing or another, crisis prone, good organizational skills, can help others develop their skills or potentials.

Famous people with Pluto in the Sixth House or in Virgo:
Marie Antoinette
Nikolai Lenin
John Dean
Sigmund Freud
Elizabeth B. Browning

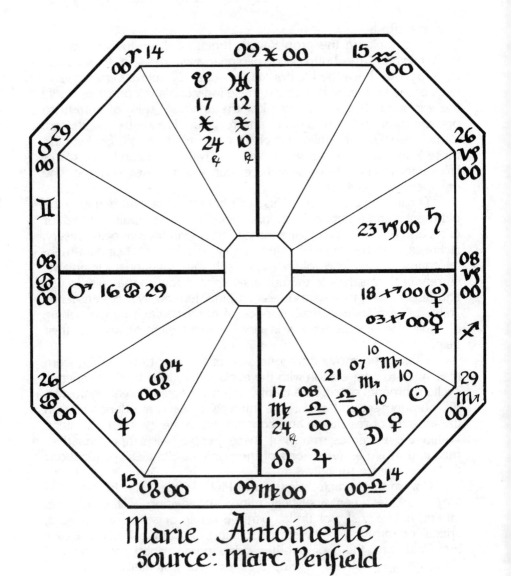

Marie Antoinette
source: Marc Penfield

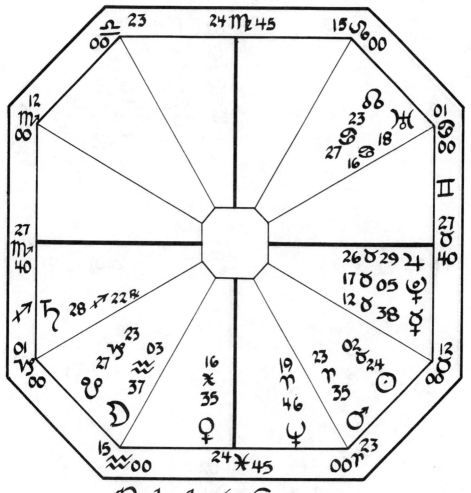

Nikolai Lenin
Source: Marc Penfield

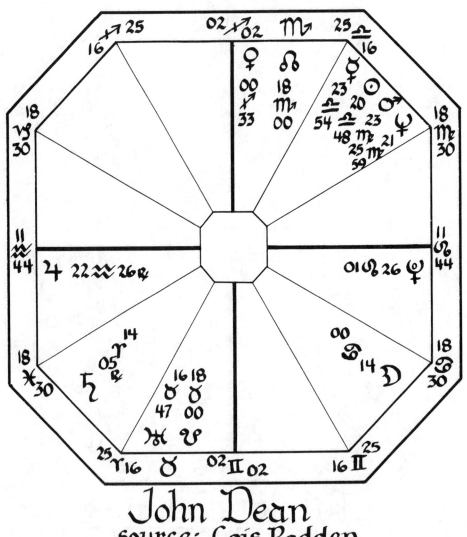

John Dean
source: Lois Rodden

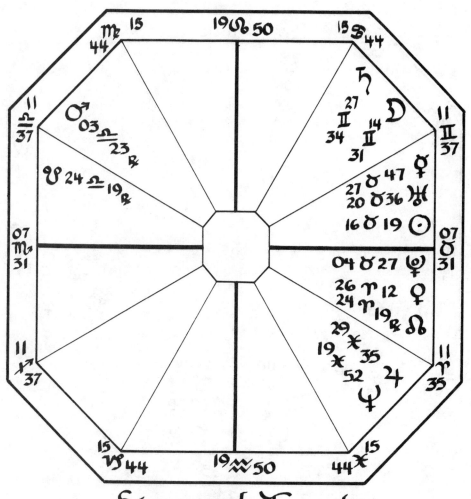

Sigmund Freud
Source: Lois Rodden

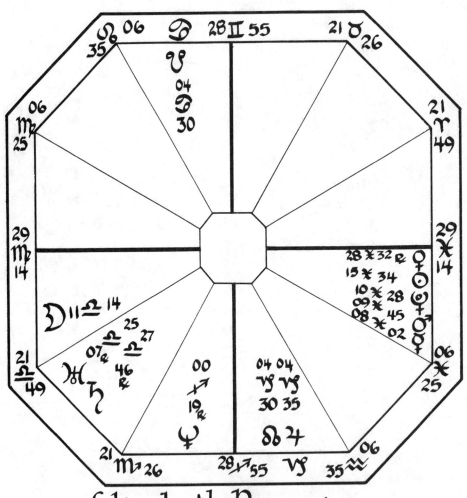

Elizabeth Browning
Source: Marc Penfield

PLUTO IN THE SEVENTH HOUSE OR LIBRA

As in all natural cardinal houses, Pluto in the Seventh House or in Libra demonstrates that a new evolutionary cycle has been and is still underway. These individuals have been learning how to develop an objective awareness of themselves and to understand their sense of individuality in a social context. In addition, they have been learning lessons of equality, relativity, and of listening. They have been learning how to listen to another's reality as it exists for them. In so doing, they have been learning how to give in relation to what others need according to that reality. Within these lessons, they have been learning how to participate in relationships as an equal.

In order to experience this evolutionary intent, those with Pluto in the Seventh House necessarily have had to be involved in a wide variety of relationships. Through involvements with a variety of people comes exposure to many different value systems, intellectual systems, emotional patterns and spiritual beliefs. Through this process these individuals have been learning how to understand their own intrinsic individuality through comparison or evaluation of the relationships that they form. In other words, it is through comparison with others that they become aware of who they are as unique individuals. In this way, these people have been learning, on a preliminary basis, the evolutionary lesson of objectivity versus subjectivity.

A fundamental lesson because of this prior evolutionary intent, is how to participate in relationships. From the First through Sixth Houses, Aries through Virgo, the focus was on the subjective development of the individual, which necessitated a self-centered orientation to life. The egocentric seed of individuality was planted in Aries. From Aries through Leo, the First House through the Fifth House, this seed progressively expanded and grew in ever widening circles of development. In Leo and the Fifth House the egocentric seed had blossomed into the creative self-actualization of intrinsic individuality. In the Sixth House and Virgo, the archetypical need was to experience essential aloneness, and to purge or purify the egocentric balloon of self-importance. In addition, it was necessary to learn the practical skills or techniques that allowed for an appropriate harnessing, focusing and channeling of the inherent work capacities of the individual to serve the needs of the culture, tribe, or society in a practical way. This archetype produced an inverted pyramid effect upon the subjective development of the individual, purging the ego of delusions of grandeur and promoting essential humility.

In the Seventh House and in Libra, the potentially purified individual is now ready to relate with others on an equal basis, rather

131

than on a subservient or dominant basis. Those with Pluto in the Seventh House will have felt in the past, and will feel coming into this life, a compelling need to be involved in relationships. This compelling need is rooted in the desire to complete the self through relations with others and to unite one's life with another's. Because of this prior evolutionary intent, the unconscious emotional security structure of these individuals is rooted in their need to be involved in relationships in order to feel fulfilled and complete.

Because Pluto in the Seventh House or in Libra represents a relatively new evolutionary cycle, these individuals have literally been learning the ongoing lesson of how to be involved in relationships with others on an equal basis. Most of these individuals have not learned how to do this. On the one hand, these individuals will need to take in information and differing points of view from many diverse types of people. This situation sets up the need to listen objectively to others in order to determine how to relate with them. By exposing themselves to the diversity of others, these individuals have been learning to understand objectively the relativity of human nature.

A potential problem that this prior life pattern promotes is one wherein the individual becomes overly involved with too many relationships, and takes in too many differing points of view and values. Becoming overly involved in this way, the individual can lose sight of his or her own individuality; he or she can become overly absorbed in the differing realities of others. When this occurs, the Seventh House Pluto individual can become compulsively dependent on the opinions, advice, or knowledge of others as a means of determining who they are, or what they should or should not be doing with their lives. The dependency is put in motion as soon as the individual loses sight of his or her own individuality.

In this condition, these individuals cannot relate to themselves without relating to others. They are not comfortable unless others are around. In addition, once the individual loses sight of his or her own individuality, the stage is set wherein the individual will be subconsciously attracted to very powerful individuals who appear to be very strong, stable, sure of themselves, and who have the ability (and perhaps the need) to guide or control the formation and development of the Seventh House Pluto person.

In this situation, the individual has reached an extreme imbalance in which they are utterly dependent on the partner or people they are involved with. In the worst cases, this situation promotes the alter-ego effect wherein the Seventh House Pluto person is promoting or fulfilling the needs of another to the exclusion of their own needs. The other person has manipulated the dynamics of the relationship in such a way as to make the Seventh House Pluto individual feel that their needs are being met by becoming an alter ego to the other. In effect, the Seventh House Pluto individual is being dominated by the needs of the other in such a way as to feel that those needs, desires, opinions, values and beliefs are their own. In this way, they have become dependent on the other to have their needs, desires, ideas, values and beliefs validated, defined and fulfilled by the other. The Seventh House Pluto individual has simply become an extension of the other's reality and identity — an alter ego.

A classic example of this phenomena is found with Richard Nixon. Nixon has the South Node in Libra relative to a Tenth House Pluto. The ruler of the South Node is Venus in Pisces in the Sixth House. Nixon collected subservient co-workers who were alter ego extensions of his own values, ideas and beliefs. Because they were extensions, these individuals also went down with Nixon and the Watergate scandal.

On the other hand, this same evolutionary dynamic of listening, observing, and identifying the reality of another to give to them what they need according to the other's reality, sets up another potentially extreme way to behave in relationship to others. In this extreme, the Seventh House Pluto individual becomes the dominant person. By listening intently, they can make others feel that they understand them, or are concerned about them, because of the ability to "play back" the other's reality to them. In effect, the Seventh House Pluto person is unconsciously attracting those who need counsel, advice, companionship, or love.

This dynamic can lead to a position of emotional control within the relationship if the Seventh House Pluto person compulsively manipulates the relationship to make the other person always feel that they need this individual in their lives to be OK, to be secure, to be loved. In this situation, the Seventh House Pluto individual will impose his or her own values, beliefs, ideas and nature on the other in such a way as to make them feel as though those values, ideas, beliefs and nature are their own. Again, this kind of relationship is in a state of extreme imbalance.

In both cases, the lessons to be mastered involve objectivity, listening and giving, and recognizing the relativity of individual needs and human nature. In both cases, the person is learning socialization. In both cases, equality and balance were not learned in the past.

Since the evolutionary intent has also been to learn lessons of equality and balance in relationships, the Seventh House Pluto individual commonly has built up karmic problems associated with a variety of people. In either one of the two imbalanced conditions mentioned above, the individual would necessarily have had to recoil or react to the imbalance through evolutionary necessity. In both situations des-cribed above, this reaction can manifest itself in one of three ways:

1. The Seventh House Pluto individual leaves the partner because the needs that created the relationship have been fulfilled. This situation sets up new needs, and therefore a desire for a new relationship. The process of leaving is not necessarily positive or easy. Often, the partner does not feel that the relationship should end, especially for the reasons that the individual claims. The relationship ends with the partner feeling unresolved as to why it ended. This unresolved condition sets up the karma to be fulfilled with that person at some other time.

2. The partner leaves because they do not feel that there is any more need for the relationship, or that they are not getting out of the relationship what they are putting into it. Their needs have been fulfilled. The Seventh House Pluto person does not feel this. *They* feel unresolved about why the relationship has ended. This situation also sets up the karma to be fulfilled with that person at some other time.

3. Both partners become mutually dependent on each other in whatever way it is defined relative to the two extremes mentioned above. The degree of dependency becomes so extreme that further growth for each individual cannot occur. In this situation the roles within the two extremes can be cyclically exchanged within the relationship, or they can remain polarized within the fixed positions of the extremes: one partner playing the dominant role, the other playing the subservient and alter ego role. In this situation one of the partners may be forcefully removed from the relationship, some-times meaning death. The enforced removal creates growth for each through the intense pain, remorse and reflection upon the nature of their individual lives, and the nature of their relationship. Neither one was consciously ready for the relationship to end. Thus, unresolved, they will find each other again at some other time.

These reactions to the imbalance of relationships are to enforce the lessons of equality, balance, and relativity. A common astrologi-cal misconception about Libra and the Seventh House is that it is balanced by nature. The fact is, the Seventh House or Libra means *learning* balance. Very commonly, those with Pluto in the Seventh House have approached relationships in extreme ways in the past. In one relationship they will be the dominant partner, and in another they will be the needy, or alter ego person. And, of course, some of these individuals have been in relationships in which the roles of being in control, and being controlled, have been interchanged within the same relationship.

Any kind of imbalance within a relationship will produce necessary confrontations, either between the partners, or internally within the individual. The internal confrontations are manifested in order to force the individual to examine the nature of every difficulty or problem within him or herself, the relationship, and/or the partner. The nature of the difficulty or problem will always be linked to a need that is not being met. External confrontation between the partners will occur for the same reason. The essence of this evolutionary lesson is to learn how to give to another what they need and in so doing to have one's own needs met. Yet, because this evolutionary impulse is relatively new, the majority of these individuals do not know how to relate to themselves or others in this way. This produces not only the problem of imbalance, but also the phenomenon of conditional giving, or conditional love. Needs are linked to expectations, and expectations are projected upon a partner, or expectations are projected upon the individual by the partner. When the projected needs are not successfully met by another, then the withholding of love or giving can occur.

Because this evolutionary impulse is also teaching the individual how to receive love or giving from others, they may not be able to accept or recognize what comes their way because of emotional distortions that are linked to the problems associated with the unmet expectations or needs. Similarly, the individual may desire to give to another, yet the other may not receive, recognize, or accept the giving due to their own emotional distortions linked to similar projected expectations or needs not previously met.

This evolutionary dilemma is necessary in order for these individuals to learn how to relate to themselves and others in a balanced and equal way. The confrontations, the conditional love, the leaving or being left by another, the shock of losing a partner through death, the alternating of dominant and subservient roles within the relationship, relationships that are formed upon the fixed roles of dominance and subservience, all occur to enforce the lessons of balance and equality, objectivity and understanding one's individuality in a social context.

These situations also occur to teach the individual what their essential needs are in a relationship, who they are because of relationships, and to develop social values regarding how people should relate to each other. These situations occur to enforce the lesson of minimizing dependency on another person, and to overcome the tendency to make another person the god of one's reality, or to allow oneself to be made into a god by the partner.

These individuals will finally come to the realization that they need to be needed, and they will realize how this need controls the formation or dynamics of their relationships. Recognizing this dynamic they will also understand their compelling need to give to others, and to be given to. The ongoing evolutionary lesson will teach them when to give and when not to, and when giving, what to give, and when not giving, why they should not give. They can also realize that by appearing not to give in certain situations, they are actually practicing a form of supreme giving. In other words, in certain conditions these individuals must learn how to withhold giving to others. The nature of these conditions will be when another keeps coming back for more without applying or working with what was given before. By learning how *not* to give in these conditions, these individuals will learn how to stop promoting dependencies upon themselves by others. This is utterly necessary from an evolutionary point of view. In the very same way, these individuals can learn how to receive that which is being given to them without becoming dependent on those that are giving.

By examining the specific karmic/evolutionary signature in each chart with Pluto in the Seventh House we can determine how the individual has approached relationships before, and what kinds of people the individual has been attracted to — and for what reasons. The natural evolutionary condition of the individual will supply additional information concerning the types and approach to relationships.

As an example, let's put Pluto in Virgo in the Seventh House, South Node in Gemini in the Fourth House conjunct Venus retro-

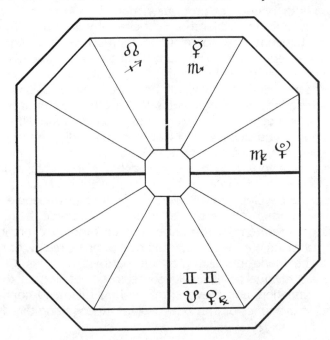

grade, the ruler, Mercury, in Scorpio in the Ninth House. This individual is in the herd state yet showing signs of individuation. In this
situation, the individual would most likely have come into this life
with displaced emotional needs based on a series of prior lives in
which the individual experienced emotional rejection by one or both
of the parents. With the ruler of the South Node in the Ninth House
in Scorpio, one or both of the parents was physically as well as
emotionally distant. This distance was necessary so that the individual
could learn the Fourth House lessons of inner security. Yet, the
individual interpreted this distance as emotional rejection, which promoted a negative self-image. The individual would feel that he or she
was not given to in the ways that they should have been. Thus, as an
adult, this individual would approach relationships with all the displaced emotions intact. He or she would be seeking information,
knowledge, healing, and perspectives from others. The individual
would expect others to supply these needs and desires. Subconsciously, the individual would attract others whose nature or personality mirrored or reflected one or both of the parents in some
way. Thus, the individual would approach relationships in a dependent and needy way. He or she would expect to be given to.

Yet, with Venus retrograde square Pluto and the South Node, the
individual would continually be rebuffed by others, which would have
the effect of turning the individual in upon him or herself in order to
recover the 'skipped steps' of emotional self-reliance and inner security.
In this case, the relationship that needs to be formed is with him or
herself (Venus retrograde conjunct the South Node in the Fourth
House square Pluto in Virgo in the Seventh House). By experiencing
an essential aloneness, and the crisis of a lack of relationships in
early and adult life, he or she would have to analyze the inner
dynamics and reasons for this experience. With Mercury, the ruler of
the South Node, in Scorpio in the Ninth House, the individual would
want to know why he or she was experiencing these life conditions.
With respect to Pluto in Virgo in the Seventh House, the individual
may seek out professional therapists in order to answer these questions. Yet, because there is evolutionary movement toward the individuated state, these professional answers may be seen as lacking
in the depth that the individual needs — they could feel wrong or
incomplete in some way. Again, the individual is thrown back in
upon him or herself.

As this process continues, the individual will one day realize that
the answers are within, and not based on accepted, prevailing, or
common intellectual or psychological cliches and theories. The
answers will be based on circumstances and dynamics that are

unique to the individual from a karmic and evolutionary point of view. Thus, the individual would learn to observe objectively and acknowledge the rejection of others as "signs" (Mercury in the Ninth House) that continually point to the need for an inner relationship with him or herself so that inner security could develop. In so doing, the individual would learn how to fulfill his or her own needs. As this transformation took hold, the individual would approach the arena of relationships in a totally different way. The individual would approach them in a self-secure way, and with a positive self-image. The individual would learn to be involved in relationships because he or she wanted to be, not because there was a need to be. In this life, this process and metamorphosis will be helped through Pluto's polarity into the First House and Pisces, the polarity sign of Virgo. The North Node in Sagittarius in the Tenth House will act as a mode of operation to help fulfill this metamorphosis and evolutionary need.

The polarity point is the First House or Aries. The evolutionary intent for this life is to teach the individual how to initiate his or her own life directions or decisions without depending on the opinions, advice, or consent of another. In the same way, this intent is teaching the individual to not control, interfere with, define, or block the partner's need to initiate directions or decisions. In effect, this evolutionary intent is teaching the individual to strike out on his or her own in order to fulfill his or her own destiny, and to minimize dependency on another in order to do so.

This polarity point does not imply that the individual should become a loner, or live without a close relationship. It does intend that the individual learn how to answer or supply his or her own needs in order to minimize expectations of another. It does intend that the individual learn how to be alone, and learn how to make his or her own decisions. It does intend to teach the individual to initiate their own self-motivated actions in order to develop more fully their individuality.

This evolutionary intent demands that the individual learn how to develop and balance the need for time alone with the need for time spent with others. In addition, the evolutionary intent demands that these individuals learn how to conduct relationships in a new way, i.e. not in the ways that one is "supposed" to. It does intend that the individual learn how to encourage and support the need for independence in the partner, and to select partners who have the courage to actualize their own life through their own means.

The evolutionary intent is teaching these individuals to develop more fully the lesson of giving to another so that the individual's own needs can be met tenfold. In this way, the balance of relationship will be realized.

As these evolutionary lessons are developed, the Seventh House Pluto individual will be one of the most giving of all people. He or she will be able to identify the reality as it exists for anyone with whom they interrelate. In so doing, these individuals can give to others exactly what they need. In addition, they will learn when to give, and when not to give. By developing these lessons, they will learn how to participate in relationships in a nondependent way, and will encourage independence in any partner or mate. Because of these lessons, these individuals will learn how to be inwardly balanced in any external situation in which they find themselves. They will no longer run the risk of losing themselves by trying on differing values, ideas, and beliefs that others represent. These individuals will learn to appreciate the diversity of human nature, and recognize their own individuality because of it. This realization can translate into the gift that they give others: the courage to discover and be themselves.

Common characteristics of the Pluto in the Seventh House or Libra: compulsive need to be in relationships or to relate with others in order to feel complete, the need to dominate or be dominated by another, the need to be needed by others, the need to be liked by others, will have a hypnotic effect on people, the need to solicite advise or to give advise to others.

Famous people with Pluto in the Seventh House or Libra:
(1) Alan Watts
(2) Jim Jones
(3) Dustin Hoffman
(4) Gertrude Zelle (Mata Hari)
(5) Immanuel Kant

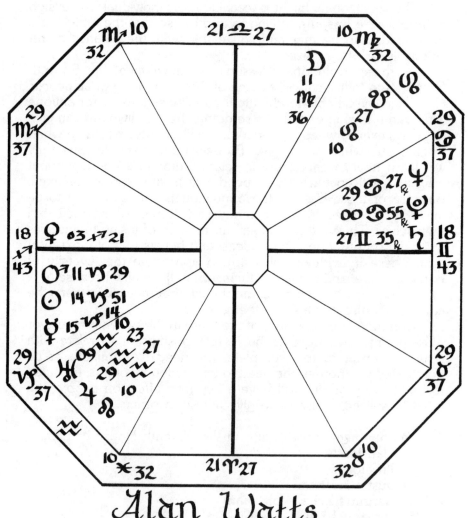

Alan Watts
Source: Lois Rodden

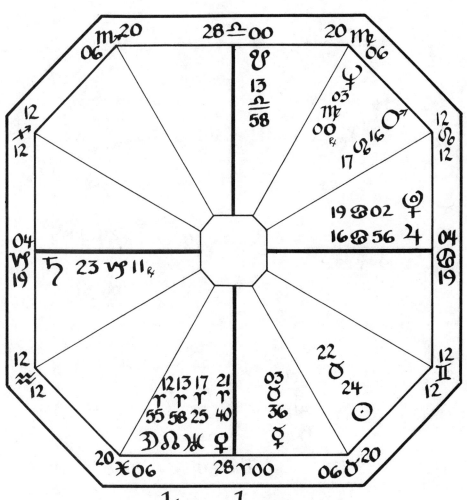

Jim Jones
source: Lois Rodden

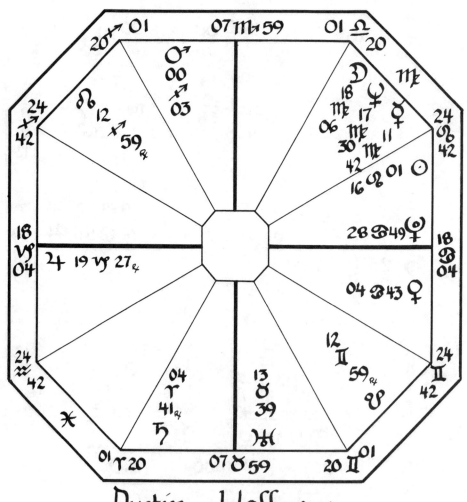

Dustin Hoffman
source: Lois Rodden

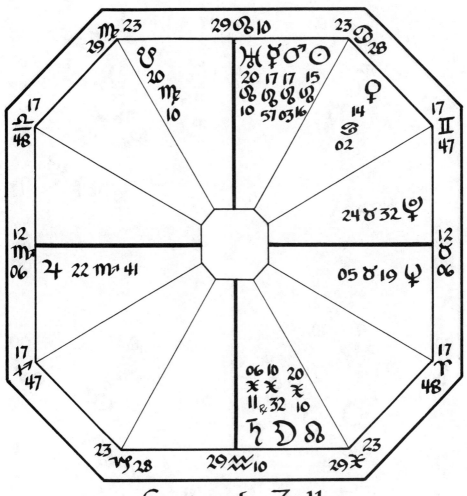

Gertrude Zelle
(Mata Hari)
Source: Marc Penfield

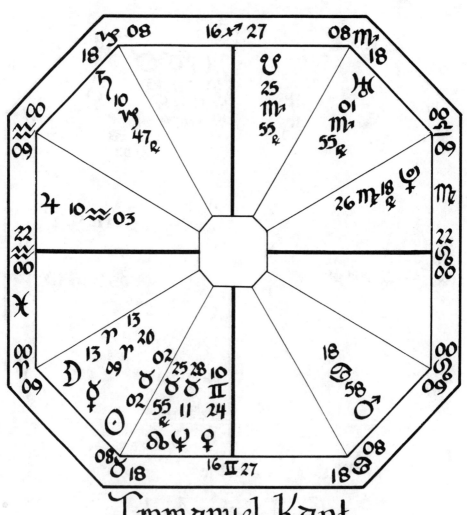

Immanuel Kant
Source: Lois Rodden

PLUTO IN THE EIGHTH HOUSE OR SCORPIO

In the broadest possible sense, Scorpio and the Eighth House are the archetypes through which all of us come face to face with the limits or parameters of who we are, as opposed to who we are not, and cannot be. We each have a specific personality or nature that is unique. Our personality is a composite of many dynamics — intellec-tual, emotional, physical and spiritual. The interaction of these dynamics determines our nature. Each individual experiences life and reality in his or her own unique way because of the specific orientation of his or her personality. Thus, we all experience the limitations or parameters of ourselves, as opposed to that which we are not, through the archetype of Scorpio and the Eighth House.

Because the essential evolutionary impulse is to grow or evolve, individuals with Pluto in the Eighth House have necessarily desired to transform their personal limitations by forming relationships to anything that symbolized what they needed or desired in order to evolve beyond their personal limitations. This process implies something that they are not, or do not already have. Thus, these individuals have experienced power and powerlessness as a result.

The experience of power comes from that which these individuals already are — what they know and have experienced. The experience of powerlessness comes from that which they are not — what they do not know and have not experienced.

The prior evolutionary impulse of Pluto in the Eighth House demanded that these individuals experience their limitations while simultaneously metamorphosing those limitations to attain greater levels of personal growth. In other words, they cannot simply acknowledge the limitations and sit back and say "Okay, that's it folks, no more".

This evolutionary impulse has enforced the awareness of the larger forces that exist in the universe, forces or power emanating from outside the individual. Evolutionary necessity demanded that these individuals merge with other sources of power in order to effect metamorphosis beyond their personal limitations. In other words, these individuals have learned how to absorb the power of whatever symbols of power with which they have formed a relationship into themselves. This absorption has promoted the required meta-morphosis of personal limitations in such a way that these individuals become that which they have formed a relationship to.

The specific nature of what symbolizes power to an individual, of what the individual feels that he or she needs in order to go beyond existing limitations, is relative to his or her evolutionary and

145

karmic condition. Examples of common symbols that Eighth House Pluto individuals are drawn to include death, sex, rituals of all kinds (magic, meditation, yoga, for example), experiences that are commonly considered "taboo", relationships, money, status, a system or body of knowledge, mantras, symbolic diagrams, God, Satan, and so forth. Uniting the self with the symbol promotes an alchemical fusion that permits transmutation of an existing limitation.

An intrinsic problem for Eighth House individuals with respect to their need to unite with these kinds of symbols is one of compulsion. As an example, Elvis Presley had an Eighth House Pluto retrograde in Cancer conjunct the South Node in Leo, trine Jupiter in Scorpio in the Eleventh House, and in opposition to Mercury and Venus in Capricorn in the Second House. It is well documented that Presley had a compulsive and morbid fascination with death and what it symbolized. Many Eighth House Pluto individuals have a subconscious death wish. Death is obviously a "power" or inevitable force that is larger and stronger than the individual from an esoteric point of view. We must deduce from Presley's case that he had subconscious guilt patterns that fueled his death wish and his compulsion to tempt death through drug use and his overall lifestyle.

In another example, a client of mine had Pluto in the Eighth House in Leo conjunct the South Node in the Seventh House, and in opposition to Mars in Pisces in the Second House. Her symbol to effect the metamorphosis of her personal limitations was a compulsion linked to sexual experience. As a little girl she compulsively masturbated. During masturbation she would have out-of-the-body experiences that allowed her to escape the confines of time, space and the limitations of her body. As an adult, she compulsively sought out one sexual partner after another and in her own words, was seeking "to extract the power of my partners into myself so that I could have more power myself." On another level she was trying to effect transcendence or death of the ego through sex. Yet, she was not consciously aware of this motive until it was verbalized for her during the counseling session. Once it was verbalized, she began to have personal insights that allowed her to stop the compulsive sexual experiences that had dominated her life and consciousness. Once she realized that she had primary lessons in self-reliance to learn, she began to get involved with the "cosmic lover" within herself and to study meditation, yoga and spiritual rituals that allowed for a personal cleansing and purification. She no longer desired or needed another person to be sexually involved with. To her amazement she found that the power she was seeking was within herself. Because she was in the beginning of the spiritual evolutionary state she united with spiritual symbols on all levels. This process allowed her to learn how to use her sexual energy in a spiritual way. Through masturbation and sexual union with another she learned how to bring her sexual energy up the spine to promote a conscious unity with the Source. In so doing, she effected a transcendence of her egocentric boundaries at all levels. By learning this inner self-reliance, she naturally drew partners to herself who reflected this same dynamic. The element of use, manipulation and need were utterly removed. She learned how to unite with the Source, and her own Soul, through meditation and through sexuality.

The experience of power and powerlessness, of transforming limitations through the relationships formed with the symbols of power, has cyclically promoted internal and external confrontations. These cycles of confrontation reflect the evolutionary impulse to eliminate and transform all limitations within the individual. Thus, these cycles lead into a continual death and rebirth of the individual's awareness and experience of him or herself within the totality of reality. This archetype correlates to reincarnation: the progressive need to eliminate all barriers (desires) that are preventing a direct merging with the Ultimate Source of power. Yet, because of the dual desires coexisting in the Soul, there has also been the need to stabilize and secure the boundaries of the most recent inner and outer metamorphosis. The desire to return to the Ultimate Source of power manifests itself as the need to eliminate progressively all limitations preventing a total merging with that Source. The desire for separateness manifests itself as the need to stabilize and secure the boundaries of the most recent metamorphic event in the individual's life.

The most intense conflict of the dual desires of the Soul occurs when Pluto is in the Eighth House or in Scorpio. The resulting internal and external confrontations arise because of the limitations in the boundaries that have been secured. These boundaries have become areas of stagnation that block further growth. The intensity of the confrontations has been and will be, in direct proportion to the degree of resistance that manifests itself through the desire for separateness. Resistance occurs because of the perceived threat to existing security patterns which are based on how the individual is relating to his or her self, to others, and to reality at any point in time: "This is who I am, this is what I know, this is how I have constructed, organized and interpreted myself, and life in general." Again, the existing security patterns reflected in this way are a direct result of a metamorphic event that has occurred at a point previous to the existing security patterns.

On a cyclic basis, those with Pluto in the Eighth House will experience emotional, intellectual, physical and spiritual confrontations in order to continue the journey of personal evolution. The resulting metamorphosis, in whatever way it is experienced, reflects the fact that the desire to remove limitations, preventing a total merging with the Ultimate Source of Power has the upper hand whether the individual likes it or not, whether one individual cooperates with it or not.

The function of confrontation also serves to produce inner psychoanalysis, and an understanding of why we are the way we are. This knowledge can be applied to a specific moment in an individual's life. It can also be applied in a holistic way to discover the individual's essential behavioral patterns and orientations that control his or her entire life, and it can be applied to the whole of humanity to explain human consciousness in an archetypical way, as Carl Jung attempted to do . Thus, in the broadest possible sense, the Eighth House and Scorpio are archetypes through which the human species examines and psychoanalyzes itself; why are we like this, why do we function this way, why, why, why? On an individual level, it is the need for each of us to ask the same questions. Thus, the confrontations experienced by Eighth House Pluto individuals promote personal as well as transpersonal knowledge. Because of prior-life experiences and orientations, those with Pluto in the Eighth House will come out of the womb asking why. Why life, why death, why this feeling, why this emotion, why this desire, why did this happen, why did they do that, and so forth. These individuals have been desiring to discover the "bottom line" of any situation that they have experienced and focused upon, the inner basis of why they work and function the way they do, and why others work and function the way they do.

By attempting to understand their own and others' internal psychology through prior-life efforts, Eighth House Pluto individuals commonly come into this life with a natural understanding of the human psyche. In effect, they are natural psychologists. Because these individuals have experienced so many internal and external confrontations over many lifetimes, because they have experienced so many fundamental metamorphoses, they will possess, in varying stages of development, knowledge pertaining to their own motivations, intentions, desires and emotional patterns,and why they work as they do.

The natural evolutionary condition or state of an individual will determine to what stage this knowledge has evolved. For those in the herd state, this knowledge can be limited to themselves and those closest to them. In this state, this limited knowledge is normally some form of "homespun" psychology that cannot be expanded upon or applied beyond their own immediate environment. In addition, this stage of development will reflect "conventional" psychological wisdom that is a by-product or distillation of collective "home-spun" psychology that has evolved over a length of time relative to a specific culture that is collectively defined. For those in the herd state who are moving toward the individuated state, this conventional psychology will be applied to everyone. Those that have evolved well into the individuated state will possess a knowledge of individual human psychology. These individuals will understand that people are different, and appreciate that each is constructed in their own way for their own reasons. These individuals cannot only apply this understanding to themselves, but can also apply it to others. They will necessarily question, confront and challenge conventional and accepted psychological knowledge. Those that are in the individuated state, yet moving toward the spiritual state, will possess knowledge of the essential psychological archetypes that symbolize and motivate human consciousness in general. They will understand themselves and others in this way. Individuals who have evolved into the spiritual state will possess knowledge of spiritual psychology that incorporates spiritual archetypes.

By becoming involved with others, the Eighth House Pluto individual will naturally be able to penetrate to the emotional, intellectual spiritual, or physical core of the other in order to understand who they are, how they work, and to understand the causes of problems or blocks that are preventing further growth. This understanding will reflect the individual's own evolutionary capacity and condition. This ability creates a natural magnetism that attracts others who feel that the Eighth House Pluto person can be of help in some way. Because of this, these individuals can hold power over others because others are giving them the power to do so.

Negatively expressed, this individual may begin to manipulate others in order to be in a position of power. They will do this by compulsively focusing on the weakest areas within another in order to manipulate and control them. In this way, they can make another dependent upon them. Similarly, these individuals can focus on the weakest possible link in any expression or form of knowledge that is not of their own design in order to put it down. This is done because of the subconscious fear of having their own power undermined by another point of view, or another person(s) who differ or challenge that which they are.

Positively expressed, these individuals can use their natural psychological knowledge to understand and motivate others. They can help others see and understand how and why they are the way they are, what the basis of their motivations and desires are, and help them purge the negative emotional, intellectual, spiritual, or physical patterns that are preventing further growth. In so doing, these individuals can help others free themselves from the limitations of their inner and outer realities so that they can move more freely on their own natural evolutionary journeys. They will give power to others rather than use power to control them. In addition, these individuals will have more of a willingness and ability to understand the source and nature of their own limitations. Consequently, they will be much more able and willing to understand what they must do to meta-morphose those limitations to greater levels of self-knowledge. They will naturally gravitate to specific symbols of power in order to effect this metamorphosis.

A key point to understand at this juncture is that if the Eighth House Pluto individual reacts negatively to transforming the limitations of his or her own egocentric identity and power, then the motivations, approach and the symbolic sources of power will be very different from those of the individual who has reacted in a positive way to this evolutionary necessity. Whether the individual has effected a negative or positive reaction, the actual symbols of transformative power will reflect his or her natural evolutionary condition. In addition, the actual symbols used with respect to the evolutionary condition in the past will be reflected by the house and sign position of the South Node, and the house and sign position of its planetary ruler. Let's use some case histories to illustrate this point.

Case One: This individual is in the individuated evolutionary condition. Pluto is in Leo in the Eighth House, the South Node in Aries in the Third House, and the planetary ruler, Mars, in Scorpio in the Tenth House. The symbols of transformative power for this individual were ideas that explained the nature of personal and social reality. The nature of the specific ideas to which the individual instinctively gravitated were related to black magic. The attraction to these kinds of ideas reflected the individual's desire to penetrate what society has labeled a taboo subject. In addition, the rituals of black magic reflected the individual's desire to control universal forces larger than himself. By learning and uniting with the specific methods, techniques and rituals of black magic, this individual effected a transformation of his egocentric identity: he became what he was learning through the osmosis effect. The need to control universal forces larger than himself so that he was not consumed by them reflects the fact that he had subconscious fears to that end. Thus, this individual also attempted to use "his" knowledge and power to

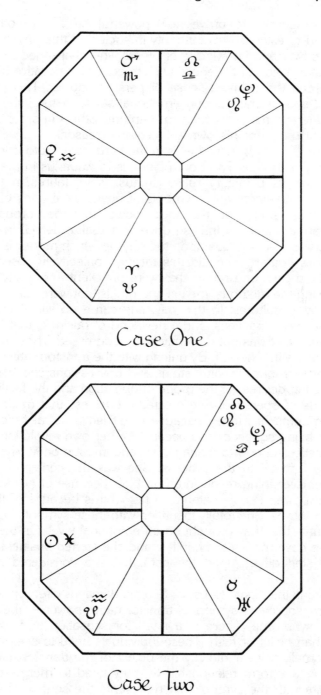

Case One

Case Two

control others, or to prove how powerful he was to others. He attempted to establish his authority in society in this way, and rejec-ted the power and knowledge of others who challenged him. Thus, the transformation of his egocentric structure was negative by nature: it reinforced the delusive sense of personal power. This individual also had an extremely strong emotional sexual nature. He used his knowledge and magical rituals to "capture" others in order to satisfy his sexual urges for his own self-centered reasons.

Case Two: This individual is in the herd state. Pluto is in Cancer in the Eighth House. The South Node is in Aquarius in the Second House, and its planetary ruler, Uranus, is in Taurus in the Fifth House. The transformative symbols of power were ones of money, status, possessions and the social power that they brought. This individual had such an intense drive to actualize herself in this way that she was able to focus upon the ways in which money is made in a business context. She did this through objectively observing the methods, means and rituals that were prevalent in her society, and by studying the methods, means and rituals that were used in history and in other cultures. In this way, she came up with "new" ideas about business practices, and a new kind of business that fulfilled a social need that was not currently being addressed. These new ideas came from within herself. By uniting with the transformative symbols of power — money, wealth, status and possessions she was able to actualize her desires via the mode of operation and the facilitation of that mode of operation. She created a business that made tremen-dous amounts of money because it fulfilled a social function that was not being supplied, and because of her own revolutionary busi-ness practices in how to manage, use and invest money. She became relatively famous in the process. She was the center of her own universe. Everything revolved around her because of her social and financial power. People catered to her whims because of their own ulterior motives. She totally identified with these symbols because of the security that they brought on an emotional level. On the positive side, she gave money to charities, and she started business schools to help start others upon the path that she considered most im-portant.

Pluto in the Eighth House or Scorpio demands direct experience of any form, source, or symbol of transformative power — the "prove it to me" syndrome. There is a need for sensation and experience rather than belief or faith. These individuals need to see, feel, taste, touch, smell, or hear directly the object of the transformative sym-bols that they form relationship to; they need to merge, penetrate and unite with the symbol that represents the source of power or object of their desire. Through direct experience they merge with the transformative symbol and become the symbol itself. Negatively, the individual will attempt to control, manipulate and use the symbol for egocentric ends — as in case one above. Positively, these individuals will allow the power of the transformative symbol to enter them, allowing for an alchemical fusion (metamorphosis) that takes place of its own accord — as in case two. Some other examples of symbols that allow for an alchemical fusion to take place are Tai-Chi, various forms of yoga, meditation, sex, merging one's resources with another, staring at a candle flame, and so forth. The symbols that promote this alchemical fusion can be almost anything. It is the person's approach, attitude, motivation, desire and the evolutionary and karmic conditions that determine the selection of the specific symbols, and the individual's positive or negative reactions to them that leads to the type of transformative experiences associated with them.

In the Seventh House and in Libra the evolutionary need was to initiate a wide variety of relationships with other people in order to learn lessons of individual relativity and objectivity. In addition it required that the individual learn lessons of individual relativity and objectivity. In addition it required that the individual learn lessons of individual and social relativity and objectivity, and how to relate with others in a variety of ways so that an inner balance and the true essence of their individuality could be realized. In this way these individuals learned who they were and who they were not. The individual also learned what his or her essential needs were in a relationship, and learned how to listen so that he or she could give to another what they needed. In the Eighth House or in Scorpio, the evolutinary intent has demanded that the individual learn how to make choices concerning whom to be involved with, and whom to commit to in a relationship. Individuals with Pluto in the Eighth House have been learning to choose from among a variety of people another individual who resonated with their own needs, desires, karma, emotions, and their Souls.

In the Seventh House and Libra, the individual was learning what he or she needed in a relationship. Individuals with Pluto in the Eighth House know what they need and have been learning to make the appropriate choices. Others who do not reflect these needs and who do not resonate with the individual will be excluded. At a core level, among other needs and karmic issues, will be the need to either seek out a relationship with another who symbolizes something that the individual desires because of a perceived need that is not being met, or to seek out an individual who is looking to the Eighth House Pluto person as the transformative symbol in his or her life. In either case, various degrees of use and manipulation will occur in order to gain that which is needed. In addition, either case implies emotional dependence.

Because of the evolutionary need to unite with symbols of transformative power, in this case another person, we also have sexual union as a function of this kind of relationship. Sexual union can create a mutual penetration and bonding of the Souls. It can create a metamorphosis of personal limitations because the two individuals have physically and symbolically penetrated each other. Is it not true that after sexual involvement with another that the person is always inside of us in some way? We take in, or take on, the individuals with whom we have shared ourselves sexually. From a spiritual or metaphysical point of view, we also take on the karma of another with whom we become sexually involved. The exchange of karma occurs through the mutual exchange of sexual fluids, and through the merging of the physical, emotional and etheric bodies. In this way, a merging of the Souls occurs. In the Second House and in Taurus, the biological instinct was procreation. The Eighth House or Scorpio is the fulfillment of that instinct through pair bonding. The Second House is the instinct and isolation of masturbation, whereas the Eighth House is sexual activity and union with another. In this way, the Eighth House (and Scorpio) is the true "marriage" house. In almost all cultures, a relationship is not considered a marriage until it is consummated.

The inherent evolutionary and karmic problem generated by the Eighth House Pluto is based on using or manipulating another in order to have one's own desires and needs met. Because of the ongoing evolutionary necessity for growth and metamorphosis, the needs and desires in the Eighth House Pluto individual also keep changing. Thus, the relationships formed in this way usually are maintained only as long as the need exists. The duration of the relationship is different in each case, different with each person. Once the need has been fulfilled, there is no further use for the relationship. It has reached a point of stagnation or nongrowth from the perspective of the Eighth House Pluto person.

At this juncture, these people face an evolutionary dilemma. On the one hand the desire and need for commitment exists. On the other hand there is the desire and need to grow. The question is: what to do? As this internal and external confrontation intensifies, the "demons of temptation" will manifest themselves within the individual — secret desires, secret attractions, phantom images and urges that would end or threaten the existing relationship if followed. This inner process naturally reflects itself outwardly in the form of sudden attractions to other people, or situations that the individual feels a desire to pursue.

These temptations can have tremendous power because they symbolize change. Often many of these people do not understand what is happening or why. The need for stability and commitment may cause them to resist temptation. This resistance can be so complete that the temptations are driven into the dark and controlled recesses of the subconscious. Yet, because there is a need to transform the stagnation that they are experiencing in their life, and in a relationship that reflects this stagnation, the demons live on; they seem to have a life of their own. The temptations pick up momentum and power in direct proportion to the resistance or unwillingness of the individual to change, to the unwillingness of a partner to change, or to the inability of the relationship to change.

What to do? One possible course of action that many Eighth House Pluto people have followed, male or female leads to what I call the "black widow" archetype. Just as the female black widow spider magnetically draws the male to her in order to be impregnated, and then kills the poor devil (of course the male black widow was following his own evolutionary role and desire), so too can these individuals use, manipulate and maintain a relationship for the duration of a need, and then move on to the next relationship.

Often, sex is used to get what they need. Sex may even be the reason for the involvement, although these individuals will resist otherwise — to themselves and others. Karmically this creates a situation wherein these individuals will attract others with whom they have been involved in another life. This situation is always unresolved because there are issues left over from a prior-life connection. Compulsion will be the basis of the attraction, and in extreme cases some form of violence can occur. In addition, a large percentage of the emotional behavior that is manifested toward each other in this life will be "irrational" behavior. The prior-life separations have not normally gone well. One or both persons will have unconscious memories of being hurt, abused, manipulated, used and then abandoned. In a typically negative Eighth House Pluto way, the need to get back at the other creates this kind of irrational behavior that is vindictive in nature normally, like rational basis exists for such behavior in this life, although the playing of the old tapes can certainly make it appear to be so.

Another course of action that Eighth House Pluto individuals have followed, male and female, leads to what I call the "die hard" archetype — maintaining a situation to the bitter end, resisting all impulses or temptations to change it and going down with the ship. This situation ultimately promotes complete emotional withdrawal, and an almost catatonic way of coping with their own life, and life within the relationship. This course of action will not only be a reflection of the evolutionary need and desires to learn commitment, but will always occur with another with whom there have been prior-life difficulties and/or leavings. Thus there is a need to fulfill this prior-life karma in this kind of way.

Another course of action, and the one to be recommended over the above, is to choose mates who are willing to evolve and change as the need presents itself. The Eighth House Pluto individual should be willing to experience life as an open-ended proposition, and to be willing to risk that which constitutes security. Those following this course of action can learn how to communicate or release that which is promoting problems, stagnation, blocks, or temptations. They can learn how to do this for themselves, and thus to do this within a relationship. This release promotes individual growth and the growth of the relationship. There are no more secrets. They can learn that some of the demons are actually angels that have assumed the form of demons because of their own fears; that some of the impulses and desires are legitimate and relevant to the individual's growth, whatever those desires are, and wherever they may lead.

In this kind of relationship, the idea or phenomenon of Soul-mates could be realized. However, strictly speaking, Soul-mates must have a spiritual foundation upon which their relationship is based and to which all that is inwardly and outwardly experienced is referred. The essence of a Soul-mate relationship is to promote the spiritual development of each person because of the relationship. The confrontations and growth pains occur, of course, yet they promote growth rather than degenerating into recriminations or vindictive, cruel or mean behavior toward one another.

Until this third course of action is adopted by the Eighth House Pluto person, the other two courses of action will promote negativity. Because one or both partners may feel that they have invested themselves totally in the other, when unexpected problems or confrontations arise, or one person leaves, negative behavior can result. Remember that these individuals can have unconscious memories of this kind of activity from other lives together, which conditions their behavior in this life. All that the Eighth House Pluto individual can act upon is a desire for revenge: an eye for an eye. In varying degrees of intensity, this one motive and desire can create the basis for the karmic connection which can exist between this person and others. Every intimate relationship that an Eighth House Pluto individual has with another will be based on a prior-life connection. This will apply to all evolutionary conditions and all possible karmic patterns and requirements.

Many Eighth House Pluto individuals have had a series of prior lives in which the rug has been pulled out from under their feet. This has occurred either through karmic retribution, or because they have overly invested themselves, and become too dependent upon, a situation or person in their lives. The degree of over-investment or dependency will determine the degree of emotional shock associated with the experience. If the individual has a karmic signature wherein they have used their sexuality to get what they needed, or to control others, then the individual may experience this type of retribution in their own lives. The worst cases might include the possibility of being raped, for example. The prior-life memory of having the rug pulled out from under their feet is responsible for most of these people coming into this life with a conditioned defensiveness, and a conditioned suspicion as to the motives and intentions of those who desire to be involved with them in some way. This dynamic creates emotional retention or withholding for fear of having the rug pulled once again. It can take an abnormally long time for the individual to really trust another or a situation in which they are involved as a result.

The polarity point is the Second House or Taurus. The evolutionary intent is one of utter self-reliance and learning how to identify one's own internal values and resources in order to sustain oneself. On this basis, these individuals are learning how to minimize compulsive dependence on anything outside of themselves. They are learning how to look within and use themselves as the symbol for their own personal transformation. The evolutionary intent demands that they progressively learn to simplify their lives — do make it more basic. Thus, it is important for these individuals to effect relative isolation from the impact of the external environment, to look within and examine the whys and wherefores of their own existence. In this way, they will learn how to identify who they essentially are versus the pieces of themselves that are actually other people with whom they have been karmically linked through relationships in this or other lives. This karmic linkage is associated with the osmosis effect of sexuality, and the uniting of themselves with others on an emotional intellectual and spiritual level. In this way, they can discover the core of themselves, and will learn how to sustain themselves by identifying their own personal resources. By so doing they will become emotionally, intellectually, spiritually, physically and sexually self-sufficient. They will learn how to participate in a committed, growth-oriented relationship in a nondependent, noncompulsive way. In this way, they will naturally attract another who is self-sustaining by nature. They will transmute the karma of manipulation because the need for growth will not be linked to or dependent upon external situations or persons. These individuals will learn how to make consistently correct choices as to what to involve and not involve themselves with. In addition, all potential misuses of power and sexuality will be completely eliminated. Walking upon the road of self-reliance, they will dance to their own tune, and not someone else's.

This evolutionary intent will allow these individuals to face the limitations implied in the totality of their natures, and make the necessary adjustments to allow for additional growth. This intent will either be desired and consciously developed, or it will be enforced through the necessary karmic blows or shocks. In the example of the individual discussed earlier who identified with the transformative power symbols of the occult and magic as a means to control forces larger than himself, and thus to control others, the North Node was in Libra in the Ninth House, the ruler, Venus, in Aquarius in the Twelfth House. In this life, he did gravitate to these old patterns again, but he experienced the karmic backlash of total rejection and isolation by others. All the magic he tried to use on others turned back upon himself. The demons of the underworld haunted his consciousness and dreams, controlled and dictated his thoughts. He heard voices calling him to do this, that, and the other thing. He ended up a paranoid schizophrenic, locked away in a mental hospital. He had no power and was at the mercy of the psychiatrists who controlled his life. This kind of emotional shock will provide the karmic balance that he needs in the long run. Conversely, the woman who had Pluto in Cancer in the Eighth House, South Node in Aquarius in the Second House, and the ruler, Uranus, in Taurus in the Fifth House, ended up leaving her businesses and schools in the hands of others. She let them run themselves because she realized their limitations concerning her own growth and needs. With the North Node in Leo in the Eighth House, and the Sun, the ruler of the North Node, in Pisces in the First House, she made a complete break and sought out volunteer work with people who were dying. She assisted them and their families and donated large sums of money to hospice programs throughout the United States. She allowed herself to be paid money from her business concerns, but only enough to get by on.

As the Eighth House Pluto individual evolves into the Second House polarity, the transformation produces individuals who are self-motivated. They will be able to encourage and motivate the growth concerns of others with whom they come into contact in an utterly nonmanipulative way, and will not allow for others to become dependent upon them. They will encourage others to strip away all the layers of conditioning that have dictated their behavior and orientation to life. In so doing, they will encourage others to realize their own individual essence and to sustain and apply themselves on that basis. Their intrinsic capacity to identify the essence of whatever issues or areas of life that they focus upon can now be used to penetrate, discover and solve the "mysteries" of life for the betterment of themselves and for others in general.

Common characteristics of Pluto in the Eighth House or in Scorpio include: intensity on all levels, deep radiating core of power, can be very fixed and stubborn, dislike of superficial relationships, can be vindictive when taken advantage of by others, strong likes and dislikes, magnetic, transforms those (for better or worse) who are involved with them, black and white without shades of gray until the necessary transformation takes place, ability to motivate others, a "why" oriented person, can be emotionally manipulative, cycles of emotional withdrawal, secretive.

Famous people with Pluto in the Eighth House or Scorpio:
    Jacqueline Kennedy Onassis
    Marlon Brando
    Winston Churchill
    Bob Dylan
    Leonardo Da Vinci

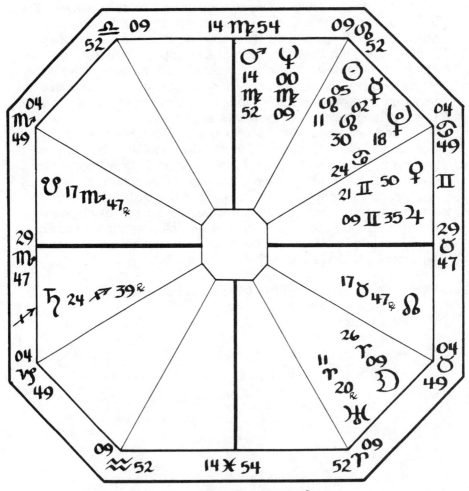

Jacqueline Kennedy Onassis
Source: Marc Penfield

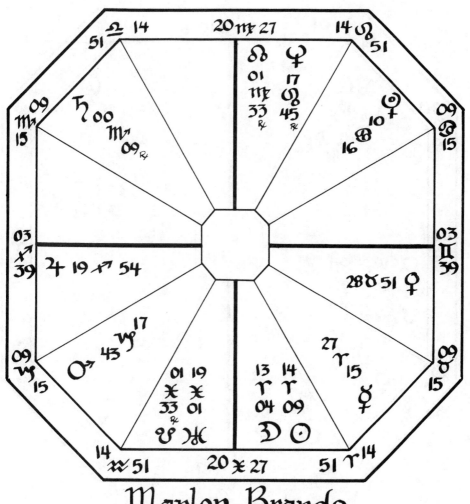

Marlon Brando
Source: Lois Rodden

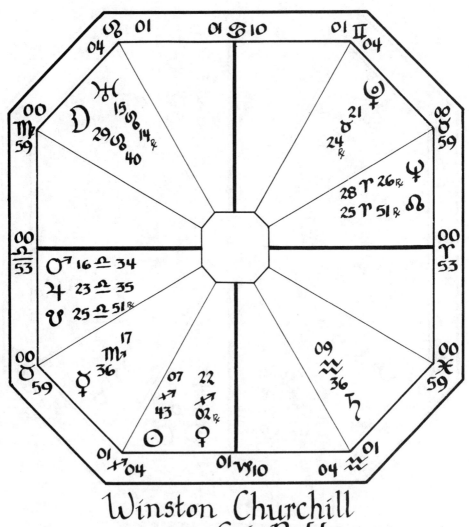

Winston Churchill
Source: Lois Rodden

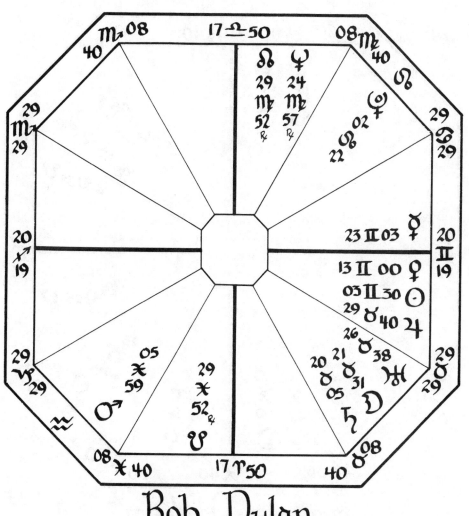

Bob Dylan
Source: Lois Rodden

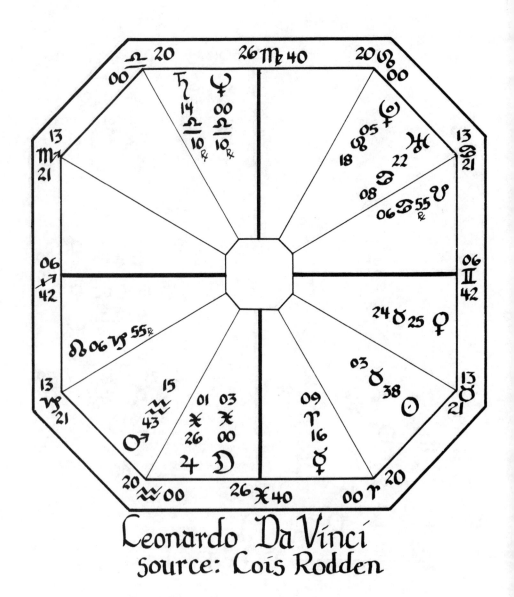

Leonardo Da Vinci
Source: Lois Rodden

PLUTO IN THE NINTH HOUSE OR SAGITTARIUS

Individuals who have Pluto in the Ninth House or Sagittarius have had the desire and evolutionary need to understand life and themselves in a cosmological, metaphysical, philosophical or religious context. In the broadest possible sense, the Ninth House and Sagittarius are the archetypes through which humanity has a need to explain its connection to the phenomenological nature of the universe and the world in which we live.

The Eighth House and Scorpio generated the awareness of "larger forces" in the universe and the world, forces that were sensed as originating outside the individual. Consequently, there was a need to form some kind of relationship to these forces in order to become identified and united with them. In effect, the relationship allowed the individual to experience these forces directly, and thus to undergo a metamorphosis of his or her personal limitations. The individual became that to which he or she formed a relationship through the osmosis effect. Because of the direct need to experience these larger forces, people and cultures have developed rituals, methods and techniques to harness, possess, control and know them. As a result, this desire and need has led into an understanding and explanation of occult or hidden knowledge. It has also led into internal psychoanalysis or psychological investigations of the inner dynamics or laws that explain who and what we are and how we work.

In the Ninth House or Sagittarius, the desire and need has been, and is, to explain one's relationship or connection to larger forces. Whereas the person with Pluto in the Eighth House or Scorpio sensed this connection, in the Ninth House or Sagittarius the connection is known. As a result, individuals with this placement have needed freedom to pursue whatever experiences are necessary to discover the knowledge that explains those larger forces.

As with Pluto in all natural fire houses, the Ninth House placement gives the individual the feeling that they have a special destiny or identity. In the Ninth House and Sagittarius, the special identity is directly linked to the individual's relationship to the cosmological, philosophical, metaphysical or religious principles that explain their connection to the universe and the world. These people recognize the phenomenological nature of the universe at large, and in contrast to those with Pluto in the Third House or Gemini, are not as focused on the immediate physical environment. They understand that, in a phenomenological sense, the world or universe simply exists; that it exists of itself beyond labels or classifications. Yet, because it exists, Ninth Pluto House individuals have desires to understand the fundamental truths that explain the existence of the world and the universe. Truth implies natural laws or principles that govern the totality of life as well as the individual's life.

On this basis, these individuals desire to be free in order to generate whatever experiences are necessary to discover the truth of their existence. They need to understand and explain their relationship to the universe via the principles founded on cosmological, metaphysical, philosophical or religious terms. As a result, many of these individuals come into this life as natural loners. Many of these individuals have traveled far and wide throughout many lifetimes in the pursuit of truth and knowledge. Many have incarnated into diverse cultures all over the planet. It is also common for many to leave a culture, society, or nation in which they felt inhibited or stifled in their pursuit of truth and knowledge. Consequently, many Ninth House Pluto individuals have and will experience a fundamental alienation from their own culture and society. Alienation occurs because most cultures, societies or nations are based on a specific religious or philosophical system, arrived at by general consensus, that serves as the framework for understanding or explaining its relationship to the world and the universe. Because the desire and drive of the Ninth House Pluto person is to search for and discover truth in an unrestricted way, alienation arises from the limitations of society. These limitations lead to external, and therefore, internal confrontations for these individuals.

The confrontations reinforce the sense of alienation from the culture or belief system into which they have been born. Given a prior-life history of incarnating into diverse cultures, many of these individuals will experience alienation because each new culture or society will be different from the one that they have just experienced. The unconscious "memories" of the most previous life experience will induce the alienation relative to the "new" culture. This produces a cultural, emotional and philosophical shock that threatens the unconscious security needs of the individual. The experience of restriction, alienation, confrontation and shock reflects the evolutionary impulse to discover the comprehensive truths, principles or laws that underlie the diverse expressions of phenomenological truth as defined by any culture, at any time.

Alienation is a necessary experience because it fosters an elimination of any barriers, be they individual or cultural, that are preventing a direct understanding of the fundamental laws and principles that govern the comprehensive and total truth of existence. Because of this evolutionary impulse, many Ninth House Pluto individuals will feel rootless, rather than identifying with any specific nation, they will consider themselves citizens of the world.

On the other hand, many of these individuals will naturally gravi-tate to certain philosophical or cultural expressions that reflect their own inner intuitive sense of how life should be explained or understood. For some this natural gravitation to a specific philosophical or cultural expression of truth is at odds with the society or cultural that they were born into in this life. The memories of the past often create a desire or need to rediscover or align themselves with a religious or philosophical tradition from another time and another place. For others the culture that they are born into will mirror perfectly the philosophical or religious tradition necessary to continue their evolu-tionary journey and to fulfill their karmic requirements. For these individuals the experience of alienation will be minimized or eliminated altogether.

Because Ninth House Pluto individuals desire to understand themselves in these ways, there is a concentrated development of the inner intuitive faculty. Because these individuals know that they are connected to larger forces beyond themselves, and because of the desire to discover the natural laws that explain this connection, they have necessarily had to develop an intuitive faculty — a "sixth sense".

Intuition is the inherent ability in each of us to "know" something without conscious reasoning. This knowing or understanding occurs without any effort on the part of the egocentric and subjective mind. Because these individuals know that they are connected to larger forces in the universe, they have "tapped in" to those forces. By intensely focusing upon those forces, they become aware of their nature and, therefore, of the laws and principles behind those forces. Laws and principles translate into concepts or conceptualizaiton of those forces. Concepts and conceptualizations translate into beliefs. A composite of concepts, laws, principles and beliefs translates into a comprehen-sive philosophical, metaphysical, cosmological or religious system.

The specific nature of mankind's belief systems are determined by individual and collective evolutionary development and geographic locality. In other words, the fundamental laws or principles are the same for all individuals or cultures, but are conceived, concep-tualized and expressed in a diversity of cultural and individual ways. This becomes clear when we observe the world's amazing array of religious and philosophical systems. Even though the basic laws and principles are the same for all peoples and cultures, the individual and cultural interpretation of those principles reflects not only diver-sity but individual and cultural necessity. Individuals with Pluto in the Ninth House or in Sagittarius will interpret these intrinsic laws and principles according to their own natural evolutionary condition and karmic necessities.

The evolutionary need to focus upon and develop the intuitive faculty in order to know the truth of reality creates many problems for the individual and for society. These problems are rooted in the two coexisting desires within the Soul. The desire for separateness results in an egocentric overidentification with a specific philosophical or religious system that reflects the needs of the society or the individual. This overidentification is based on the need for emotional, physical, spiritual and intellectual security and stability. This need for security and stability gives rise to the need for the society or the individual to defend against the intrusion of other interpretations of the same intrinsic laws and principles by other cultures or individuals. In this state of delusion and separateness, the individual or culture will either attempt to impose its system upon another or will withdraw from the threat posed by another individual or culture. The history of human experience on the planet abundantly demonstrates this fact. The recent turmoil in Ireland, the Middle East, South Africa, China's imposition on Tibet and so forth are only the latest manifestations of this sad evolutionary development.

The negative manifestation of this evolutionary condition is what I call the "Billy Graham archetype" — the need to convert others to one's own point of view. In a delusive way this need is called teaching but is really indoctrination when that which is being taught is presented with the attitude that "we are right and they are wrong; I am right and you are wrong." The stated goal of Christianity has been to "convert" those of other religions to the Christian faith. Thus we have missionaries. Of course, many other religions also seek to convert others. Sects within a religious system also occur because an individual or a group of individuals interpret the tenets of the religion in a different way than that prescribed by tradition.

Individuals with Pluto in the Ninth House or Sagittarius commonly need to convert others to their points of view because of their own need for emotional security and stability. Again, this security and stability is linked with their particular interpretation of the laws and principles that apply to all. When an individual or a culture exists in a state of separatist delusion, the phenomenon of generalizations will occur. In interpreting the laws and principles of the universe in specific ways resulting from evolutionary and karmic necessity and geographic locality, the specific interpretation is projected in a generalized way, as truths that apply to all. The principles of morality, ethics and virtues that are reflected in specific interpretations and versions of a particular philosophical or religious system are also projected as

generalized truths that apply to all. In this state the individual or culture will consider others wrong when they do not conform to their
specific version of the "truth". True teaching occurs when the individual
or culture manifests the attitude that what is being taught is only one
expression of the phenomenological truths or principles of the universe
and that these truths or principles should be considered or experienced
for oneself.

Another problem related to the development of intuition is that
the individual may not know how to communicate that which they
are intuiting through common, linear, rational language. The development of intuition is a function of the abstract mind; the ability to ponder, wonder or speculate upon the nature of existence in a cosmological
or phenomenological context — to abstract oneself from one's
immediate physical environment and to ponder one's relationship to
the universal. The power of abstraction directly relates to the development of the intuitive faculty. Thus, these individuals simply become
aware of principles and information that are not products of their
own education, deductive analysis or mental gymnastics. This is
knowledge that simply comes into the mind. Many of these individuals
may not understand why they know what they know; they just do.

A potential problem occurs when the individual attempts to
communicate that which they "just know" in a way that can be grasped
or understood by other people; to find the words that express the
intuited knowledge or that which the person already knows from
prior-life efforts. This communication problem can obviously be very
frustrating. For example, many Ninth House Pluto individuals have
had many lives in the East. The language systems of the East have a
totally different structure than those of the West. In the West a word
describes a fact, a detail or a particular object. Oriental languages
contain characters that describe whole concepts. Changing one line
within the character can alter the nuances of expression of that concept, or even change the concept entirely. Ninth House Pluto individuals
naturally think in conceptual wholes based on prior evolutionary
intent. To find the specific words to explain an intuited concept can
create many problems, especially during childhood. This can reinforce the natural sense of alienation that was discussed earlier. It can
promote withdrawal from the new life or culture, and a lifelong
search for a philosophy that feels natural and comfortable. This
search may be played out physically by traveling to many states or
countries. It can be played out intellectually through reading works of
literature or philosophy, or it can be played out by simply drifting.
These "drifters" never stay long in one place or connect in any committed way with any one person, environment or philosophical orien-

tation.

Another problem inherent in the development of intuition is the potential for experiencing visions of the future. These visions may be of a very personal nature, or they may be of a collective or planetary nature. The problem is twofold:

1. The visions may reflect deep unconscious desires that translate into "meant to be" experiences or potential experiences for the individual or others who are in close association with the individual. These visions can be of a positive or negative nature depending on the complex interactions of many karmic dynamics built up over many lifetimes. The potential problem is that the "vision" can merely be a reflection of the individual's own subconscious desires. This dynamic can create the self-fulfilled prophecy because, if the desire is strong enough, it cannot help but transpire. In this way, the individual denies responsibility for his or her own actions, or the effect of the vision on another's life, because it was "meant to be."

In other cases, an individual may truly have visions that are not based on unconscious desires. These visions simply reflect the individual's highly developed intuition into the nature of things, and the future. Again, they may not know why they know or "see" these things, they just do. Problems occur because others often do not want to know about the vision, or do not understand how the person's vision may concern them. Jesus of Nazareth had a Ninth House Pluto in Virgo. He knew the truth of his visions, yet others were confused or did not want to know about them because of the effect they would have on their own lives.

2. The individual can experience collective or planetary visions that apply to all of us. Nostradamus, who had Pluto in the Ninth House *and* Sagittarius, is a classic example. Sometimes these visions accurately forsee later events, sometimes they don't. Sometimes people listen, sometimes they don't. Some understand, some don't. Individuals with Pluto in the Ninth House or in Sagittarius are learn-ing that the future can be altered through individual and collective choices and actions. Consequently, they must accept the possibility that a vision may not come to pass, and they must learn not to become overly attached to or identified with the vision on an egocen-tric basis. Above all they should be aware that they did not "create" a vision of this kind. This kind of vision occurs because of a highly developed intuition that is tapped into universal and cosmic forces. By being unified with these forces, the osmosis effect of Pluto pro-duces the knowledge or vision. These kinds of visions occur of themselves.

In general then, individuals with Pluto in the Ninth House or Sagittarius have had the evolutionary desire to expand their horizons of personal awareness. They have needed to identify themselves with abstract principles that allow the development of a cosmological belief system that explains their relationship to the world and universe. Their emotional security is linked to these belief systems. The natural evolutionary condition of an individual, and the South Node with its planetary ruler, will relate to how this desire and need has been fulfilled in the past. Let's use some case histories to illustrate this point.

Case One: This individual has Pluto in Leo retrograde in the Ninth House. The South Node is in Capricorn in the Third House, and its planetary ruler, Saturn, is in Cancer in the Eighth House. This individual is in a spiritual evolutionary condition. He was born in Utah where the prevailing religious tradition is Mormonism. His parents were devout Mormons. His father was an elder within the Mormon church. As a boy, this individual was immersed in the Mormon point of view. His father was convinced that the doctrine of Mormonism was the only right path for everyone. He was self-righteous and intensely authoritive in his convictions. All opposing points of view were rebuked. The boy followed in the father's footsteps and did not question him. His need to link his sense of personal identity to a religious doctrine was fulfilled through his father. It reflected his own evolutionary and karmic requirements. Rather than experiencing alienation, he felt right at home in the environment of his parents and culture. As an adult, he became just as actively involved in the Mormon religion as his father had. Just as his father had defended this religious orientation against any other point of view, so did this individual. The sense of personal identity and emotional security and stability was the basis of this defense and exclusion. This man became a minister and relocated to another state where Mormonism is not widespread. In the process of trying to convert others to Mormonism, he experienced intense philosophical and intellectual confrontations with others who did not share his point of view. At this point he began to experience a sense of alienation. In addition, the intensity of the different points of view to which he was exposed in the process of trying to convert others, led to an inner questioning of his own convictions. A progressive sense of dissatisfaction with the limited and exclusionary nature of his doctrines began to occur. Insecurity and instability followed because his increasing awareness and consideration of other points of view challenged the very foundations of his sense of personal identity. The power of other points of view made him feel progressively more powerless. Confronting and questioning his father

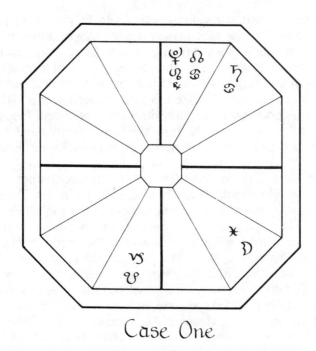

Case One

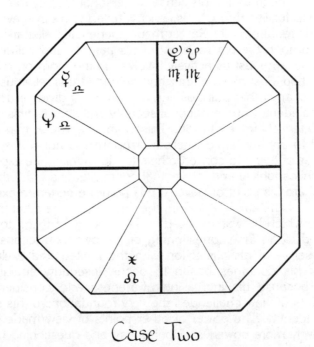

Case Two

and the elders of his religion only intensified his dissatisfaction and alienation because their answers and reactions were always the same.

   Case Two: Pluto is in Virgo in the Ninth House conjunct the South
Node in Virgo; the planetary ruler, Mercury, is in Libra in the Eleventh
House. This individual is in the preliminary stages of the individuated
evolutionary state. She was born in Sweden where the prevailing
Christian tradition is dominated by the Church of Sweden. In the
early part of her life she was exposed to the doctrines and teachings
of that Church. She found that her need to understand herself and
reality in cosmological terms was satisfied to some degree, yet, as
she became older, she felt a progressive alienation and sense of lack
in these limited doctrines. Along with a major percentage of her peer
group, she began to criticize the limitations of this doctrine and its
interpretation by the ministers of the Church. Her developing sense
of intuition knew there were different ways of viewing and understanding life in a cosmological and religious context. Speculating upon
this issue intensified her sense of alienation. The Church did not
have the answers for her or her peer group. Finally rejecting the
Church outright, she aligned herself with others who felt similarly
alienated. Together with this group of people she adopted a lifestyle
in which the philosophy, morals and ethics were based on an agnostic orientation to the world. By adopting this point of view she neither
denied nor admitted the existence of God or the rightness or wrong-
ness of the doctrines that followed from that point of view. In this way
she did not try to convert others to her point of view other than to
challenge those who did not question the certainty of their own convictions and beliefs. "How do you know?" she would ask. By adopting an agnostic philosophical and cosmological orientation, she
created maximum freedom to experience and explore whatever she
needed to in order to answer her questions. Although she was a
loner at heart, she needed to belong to a peer group who shared the
same philosophical orientations in order not to feel alone. In so
doing, she was not alone, did not feel alone, nor did she appear to
be a loner on the surface. By exposing herself to a peer group in this
way she was subjected to different opinions and values. Experiencing
lack and confusion in herself and needing to belong to something
she tried out many different ideas, beliefs, values and lifestyles. Each
one worked for a time, yet each ended up lacking the sense of cosmic meaning that she was fundamentally seeking. As she got older
the sense of alienation deepened, she felt more lonely, and criticized
herself for going along with the crowd. She did not know what to do
or where she fit in with respect to the cosmic scheme of things.
However, she remained steadfastly an agnostic who challenged others

who were not.

The polarity point is the Third House or Gemini. The evolutionary intent for this life is one wherein these individuals must learn that their version of reality, their beliefs, principles, laws, ethics and morality are relative. They are learning that the paths to truth are many, and that each person has their own path and way of discovering the truth for themselves.

This lesson occurs through necessary philosophical or intellectual confrontations with other people or through cultural alienation. By experiencing diversity or differences in this way, the fundamental security of these individuals will be challenged. Their security is linked with the conceptual framework or beliefs with which they have identified themselves — be it atheism, existentialism, any religion, agnosticism and so forth. Having their security challenged or confronted in this way, these individuals can be very resistant to other points of view. Thus, they can be very argumentative as they attempt to convert others to their point of view by exposing the weaknesses in another's argument or convictions. And yet, they are destined to draw to themselves other people who are as powerful, or more powerful, than themselves. These others will have different points of view or beliefs that show the weakness or limitations in the individual's own system of beliefs or conceptual organization.

These confrontations can also occur through others who simply do not grasp what the individual is trying to communicate. This kind of confrontation occurs in order to teach the individual how to communicate in a language that is understandable to others.

In order to do so, these individuals must study the diversity of life with respect to all the "levels" on which it exists. In studying life in this way, the individual is forced to realize the relativity of experience, of paths and of truths. In this way, the individual can learn how to communicate in a logical and succinct way in order to share with and relate to others within the context of their reality. By so doing they can allow themselves to be educated by others, and others can be educated by them. Thus the individual can learn how to expand the frontiers of their awareness in even larger ways.

Through both of these evolutionary routes, the Ninth House Pluto individual will progressively learn to understand the essential unity of all paths, of all truths. They will learn how to take many diverse facts or pieces of information, and through their inherent ability to intuitively synthesize arrive at ever greater understanding of the total nature of personal, collective and universal truths. They will thus progressively become less defensive and develop an attitude of appreciation and understanding of the relativity of truth as expressed through individuals and cultures. There will be no need to convert. In its place will be an attitude of mutual sharing, of suggestion rather than conversion, of teaching rather than indoctrination.

Many Ninth House Pluto individuals bring a natural wisdom into this life based on prior-life efforts. Many are natural teachers as a result. Yet, until they learn Third House or Gemini evolutionary lessons they will be karmically blocked from being able to express or meaningfully relate their knowledge to other individuals or to society. Some will interpret this block as a "sign" to dissociate themselves from the mainstream of cultural life. They will rationalize this "sign" into an abundance of philosophical justifications to explain their non-involvement with the world into which they are born. Others who refuse to understand the real meaning of this karmic "sign" will similarly use philosophical justifications to explain these kinds of actions. These types become the alienated drifters who forever roam the highways and alleyways searching for something that is never found. Others take refuge in their cherished belief systems and remain isolated from those who do not believe as they do. Relating only to those who share their principles, they hang out together in a state of philosophical and cosmological smugness, looking down upon others who do not share their point of view. Those with Pluto in the Ninth House who understand the real significance and meaning of the "sign" resolve to understand the evolutionary lessons at hand. According to their evolutionary capacity, they can transform our vision as to the nature of reality in ways that are as diverse as life itself.

The Mormon minister's North Node was in Cancer in the Ninth House, and its planetary ruler, the Moon, was Pisces in the Fifth House. This man finally developed an awareness that all the paths to God were legitimate. He made a complete break from Mormonism and his father because they could not accept his new attitude. He became a minister in the Unity Church through which he not only spread this kind of teaching, but also promoted the expansion of his own self-realization. His oratory inspired others because he was a living embodiment of what he was teaching. The Swedish womans' North Node was in Pisces in the Third House, and its planetary ruler, Neptune, was in Libra in the Twelfth House. She ended up moving to a Sioux Reservation in the United States. She became a nurse practitioner, helping the Native Americans through her work and adopting their values, beliefs and lifestyle. She discovered the sense of cosmic meaning that she was seeking, and found her place within the cosmic scheme of things. She also acted as an individual who bridged the cultural gap between the Sioux and the government. By committing herself to the Sioux cosmological orientation, she was able to experience the living Divinity through the techniques and methods that she was taught. As a result, she was no longer an agnostic.

Common characteristics of individuals with Pluto in the Ninth House or in Sagittarius include: a deep feeling of alienation, extremely intuitive, conceptual thinkers, can be very fixed in their beliefs, need to convert others to their point of view, value honesty, can be natural loners, philosophical, ability to laugh at oneself, can make others laugh at themselves, natural storytellers with a tendency to exaggerate, concerned with truth and not opinions, tend to be "natural" people rather than cosmopolitan and sophisticated.

Famous people with Pluto in the Ninth House or Sagittarius:
Kahlil Gibran
Michel De Nostradamus
Michaelangelo
John F. Kennedy
Jesus of Nazareth

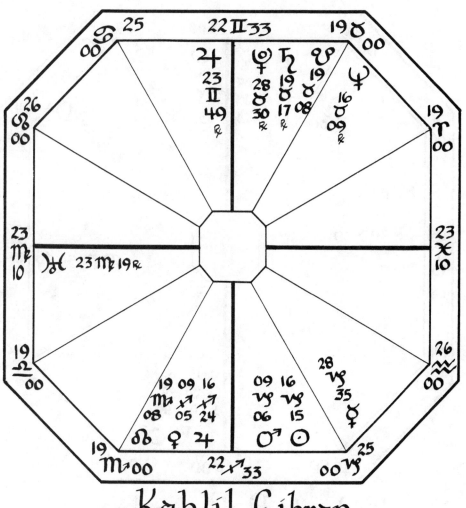

Kahlil Gibran
Source: Marc Penfield

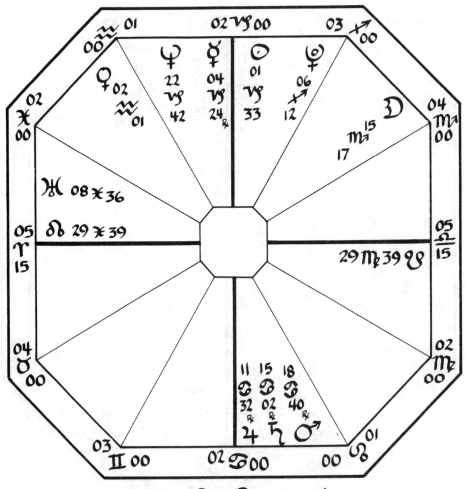

Michael De Nostradamus
Source: Marc Penfield

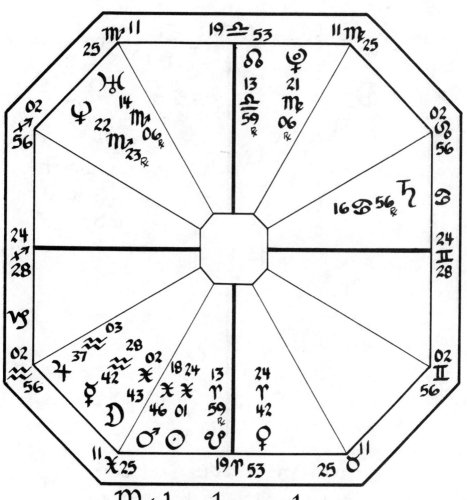

Michaelangelo
source: Lois Rodden

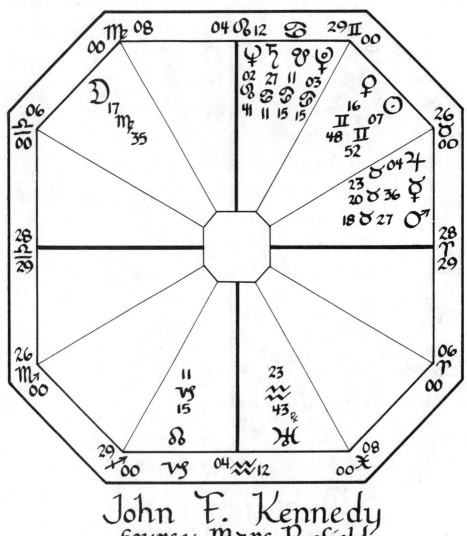

John F. Kennedy
Source: Marc Penfield

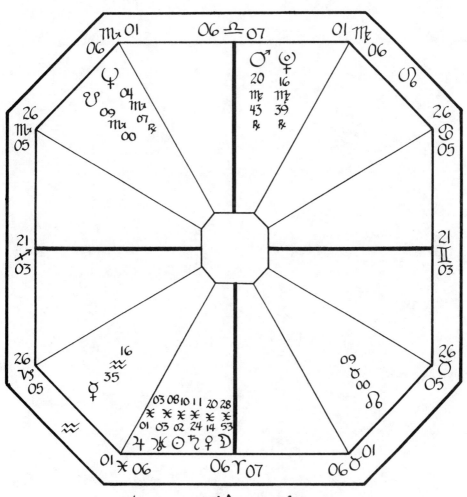

Jesus of Nazareth
(porphyry house system)
Source: Donald Jacobs
(Jacobs used equal house system)

PLUTO IN THE TENTH HOUSE OR CAPRICORN

As in all natural cardinal houses, Pluto in the Tenth House or Capricorn represents a condition wherein an entire chapter or cycle of evolutionary development has come to a close and another has just begun within the recent past. This evolutionary development can apply to countries as well as individuals.

For example, the U.S. chart has Pluto in Capricorn, South Node in Aquarius, and the North Node in Leo. The rebellion against British rule (South Node in Aquarius) was necessary in order for this country to take charge of its own destiny (North Node in Leo). In so doing, a new evolutionary cycle began (Pluto in Capricorn). The pyramid reality structure of the North Node in Leo seems to be reflected in the fact that so much of what happens in the world depends on what happens in the United States. The U.S. was destined to play the international role (South Node in Aquarius) as the "leader of the free world" (North Node in Leo with respect to Pluto in Capricorn).

Individuals who have Pluto in the Tenth House or Capricorn have been learning the evolutionary lesson of how to establish their individuality and authority within a culture or society. Any society must establish laws, regulations, customs and taboos to which each individual within the society must conform so that the society can function as an integrated and stable whole.

These laws and customs may evolve naturally over time, may be established through a consensus of the society's members or they may be imposed upon the society by those in power. In any case, a society or culture represents collective authority that dictates how individuals are expected to conduct themselves. Each nation or culture has a philosophical tradition, derived from the Ninth House or Sagittarius, that is the foundation upon which the laws, regulations, and customs are formulated. In the Tenth House or Capricorn, these philosophies become institutionalized and politicized. Examples include the philosophy of democracy and religious freedom (Ninth House) evolving into capitalism (Tenth House), or the philosophy of Marxism (Ninth House) leading to communism and atheism (Tenth House). In this manner a nation or society creates traditions and forms a national identity.

In the same way, individuals who have Pluto in the Tenth House or Capricorn have needed to learn how to link or integrate their own philosophy and identity within the framework of a society, culture or nation. This evolutionary necessity and development represents the progressive socialization of the individual which began in Libra or the Seventh House. Just as a country has borders that define its relationship

183

or position with respect to other countries, so too does the individual have a position, place or function within his or her country. Each individual has boundaries, parameters or limitations relative to his or her social position or function within a society. As a result, Tenth House Pluto individuals have had to learn, and will continue to learn, how to define their individuality through their social position or function. In order for this to occur, they must learn how the society works; if I have a desire to do this or that, to become this or that, how does society allow me to accomplish my goals?

By learning to operate within society, these individuals have also been learning related lessons in discipline, self-determination and submission to a higher authority. Because they have been learning how to establish that which characterizes their individuality, i.e. values, beliefs, desires and needs, and to link their individual purpose to a social purpose, they have also been learning lessons in social and emotional maturity.

Because each nation has its own vibration that is created through its laws, customs and taboos, each of us is born into a nation that reflects our karmic and evolutionary needs. Just as the Fourth House and Cancer described our experiences with our parents and early environment, the Tenth House and Capricorn describes our "extended family" environment; the country or culture in which we grow up. Just as our self-image is conditioned by our parents so much of our sense of social identity is conditioned by the nation in which we grow up.

Tenth House Pluto individuals, then, have had, and will have, an intensified evolutionary need that demands they learn to become socially responsible people through the social functions that they are karmically destined to perform. The natural evolutionary condition and karmic lines of development will correlate to those functions. These individuals will be born into countries that best reflect the ways in which they can learn these lessons.

A majority of people with Pluto in the Tenth House or Capricorn have desired positions of social power or status. Many have occupied positions of power in past incarnations. Power, in this context, is relative and based on evolutionary capacity and karmic dynamics. In order to accomplish their ends, these individuals thoroughly learned how the system works using Pluto's innate ability to penetrate to the basic structure or essence of any subject. In the Tenth House, this means society or social systems.

Many people with Pluto in the Tenth House or in Capricorn have learned to manipulate the system to their own advantage. They have done so in order to realize their ambitions: to get to the top. These individuals often believe that their ends justify the means. In extreme cases, this dynamic can lead to blind ambition. An example of this extreme is found in former President Richard Nixon who has Pluto in the Tenth House.

The desire for social power or status also reflects the fact that these individuals need to be in control of their position within society, rather than allowing society to control them. Pluto in the Tenth House has made the individual aware of his or her power and powerlessness and what symbolizes power and powerlessness in a societal context. The nature of a society determines the ways and means through which social position is achieved. Thus, if the system is corrupt, unjust or manipulative, the individual who seeks power and position within that system is likely to do so in a corrupt, unjust or manipulative way. In so doing, they claim that "This is just the way it works" and relieve themselves of any personal responsibility.

The potential karmic and evolutionary problems that this can lead to are twofold: 1. becoming overly-identified with one's career, position or status, and 2. manipulating and using others for one's own ends. Both can lead to abuse of power, especially if the person's entire sense of emotional and physical security is linked to his or her social power, career or status.

When this occurs such individuals, or even entire nations, may use all means at their disposal to defend themselves against any usurping of their power, position or status. The effort to maintain power or position at any cost can assume very ugly forms. Nixon's response to the Watergate investigation is an example of this. A recent example of a nation acting under the banner of "national interests" was the U.S. invasion of Grenada. A grosser example would be the invasion of Czechoslovakia by the Soviet Union. The rationales used to justify such actions are always self-serving and pious.

The history of humankind is a sad testimony to these facts. War, invasions and defense against invasion seem to be the rule rather than the exception in human history. The struggle among nations to dominate or be dominated predates the beginnings of history itself.

Yet another evolutionary lesson of the Tenth House Pluto is to learn how to be responsible for one's own actions. Individuals who have manipulated others to achieve their own ends will come into this life with subconscious guilt patterns as a result of prior errors. The same principle applies to nations. Even today the German people feel guilt in their national Soul because of the holocaust created by Adolf Hitler and the Nazis. Hitler had an Eighth House Pluto in Gemini conjunct Neptune. His South Node was in Capricorn in the Third House conjunct Jupiter and the Moon. The ruler of the South Node, Saturn, was in the Tenth House in Leo. The North Node was in Cancer in the Ninth House, ruled by the Capricorn Moon conjunct Jupiter and the South Node. Later, I will use Hitler's chart to illustrate the points covered in the first two chapters.

The guilt occurs because the standards of conduct inherent in Pluto are intensified in the Tenth House or Capricorn. No matter how it is rationalized, the abuse of power is not ultimately right. All of us have a deep inner sense of what is ultimately right and wrong. We may not know exactly why something is right or wrong; our judgment derives from this inner sense or intuition.

Commonly, then, many Tenth House Pluto individuals will come into this life with a deep sense of inherent guilt that cannot be rationally explained by current life experiences. This situation will only occur when there have been negative or misguided behavioral responses to the evolutionary impulse of the Tenth House Pluto. This guilt can have a very profound inhibiting effect in the current incarnation, so much so, that these individuals often feel that they do not deserve anything beyond what they already have. They feel resigned to their lot in life. In this way, they are punishing themselves for prior "mistakes" and attempting to atone for prior sins. The greatest frustration occurs because many do not know why they find themselves in this condition. Karmically, many will be blocked by society from fulfilling their goals. The doors seem to be closed. Others block themselves, and close their own doors. This condition enforces reflection, through which the individual can become aware of the inner motivations or dynamics that created this situation. Reflection can lead to deep depression and feelings of futility, of an utter lack of power, an utter lack of being in control of life. Often these individuals will draw to themselves a family situation in which one or both of the parents will be stern disciplinarians, imposing rigid standards of conduct on these individuals as children. Or they may have a parent who leaves the scene, or rejects the responsibility of parenting. This situation leads the individual early on to reflect upon the actions of one or both parents, and reflects their own evolutionary and karmic need to judge themselves relative to their own standards of conduct. Often these individuals have been heavily judgemental in the past, judging others against whatever standards of conduct and behavior they identified with at the time.

A primary lesson of the Tenth House Pluto has also been self-determination and self-discipline. Self-determination means identifying one's goals, abilities, capacities and desires and actualizing them through one's own efforts. The need is to identify one's function within society, and to actualize the potential of that function through the ways and means prescribed by society. This process requires discipline and commitment to a sustained effort.

Many Tenth House Pluto individuals have not learned this lesson because the evolutionary impulse is quite new. Yet, their pre-existing patterns of identity association will revolve around the need to identify with a career as a primary focus of their sense of individual completion, meaning, and security. Many have settled for less in their most recent prior lives, either because they did not learn the appropriate procedures that would have led to the realization of their ambitions, or they refused to make the necessary effort, or both. Coming into this life, these lessons will remain.

Some individuals who have had such prior-life experiences will go through cycles of pessimism, depression, bitterness, spitefulness and jealousy of those who are in positions of authority in this life. They may be unable to define or accept the limits of their social capacity and be unable to follow the prescribed ways and means of actualizing what capacity they do have. Guilt can occur through the inner awareness of their failure to utilize their potential or guilt can be projected outward if they blame the stifled conditions of their lives on society or on others in general.

In truth, the responsibility for such conditions is their own. Often, the parents reinforce their guilt because they judge their childrens' shortcomings against their own standards of conduct.

On a more positive level, those who have mastered the lessons of self-determination, self-discipline and the ways and means that a society prescribes to realize ambition, capacity and social function, have held meaningful social positions in other lives. Because they have learned these lessons before, these individuals can actualize the social function for which they are evolutionarily and karmically destined with little effort. The roles that they have played and can play, large or small, commonly have a positive and transformative impact either within their specific area of work or within society as a whole. Those who have learned to conduct themselves in relation to transcultural and timeless standards of conduct, or have at least determined to attempt to conduct themselves in this way, can play roles that inspire others to improve their own lives. This impact can be small or large depending on the evolutionary and karmic signature of the individual. In certain cases, this impact has transcended the borders of the individual's native country and touched others in other lands. A classic example is Paramhansa Yogananda. Yogananda had Pluto conjunct Neptune in Gemini in the Tenth House, South Node in Scorpio in the Third House conjunct Uranus, North Node in

Taurus in the Ninth House, and its planetary ruler, Venus, in Sagittarius in the Fourth House conjunct Mercury. The world role that he was destined to play took him to many lands in order to bridge cultural gaps through the teachings that he expressed. With an opposition from Pluto to Venus, it is interesting to note that he had a deep personal desire to seclude himself in the Himalayas that went unresolved in his lifetime. He made the statement that after leaving this life he would reincarnate hundreds of years later to fulfill that unresolved desire. Another example of a Tenth House Pluto individual whose life and work transcended national borders, albeit on a different level, is Agatha Christie. Christie also had Pluto conjunct Neptune in the Tenth House with the South Node in Sagittarius in the Fourth House conjunct Mars. The planetary ruler of the North Node, Jupiter, was in Aquarius retrograde in the Fifth House. The ruler of the North Node in Gemini in the Tenth House, Mercury, was in Libra in the Second House. Many of her mystery novels had international settings and her books have been published in dozens of languages all over the world. Intrinsic to a mystery is the theme of right and wrong and the assumption that what is just and right will prevail in the end. Christie's writings expressed such standards of right and wrong.

Those individuals who have made an effort to learn the required evolutionary and karmic lessons in the past will express natural leadership qualities at whatever evolutionary level to which they have evolved. They will be the embodiment of self-determination and can act as models for others who wish to motivate themselves in the same way. These individuals will understand that they are responsible for their own actions, and will not use devious means of achieving their ambitions. Commonly, they will feel an intense contempt and dislike for those who resort to devious ways of realizing their goals, and are not afraid to confront those who do use these tactics. The Tenth House Pluto person thus exposes others to their motivations that make them operate as they do. In so doing, they have the potential to influence others toward a higher standard of conduct.

These individuals have commonly had a parent or parents who support, encourage and understand what they must do with their lives. Sometimes one parent is supportive and the other is not. Sometimes the one that isn't eventually comes around to see that the path that the individual must travel is right for them, even if it does not conform to the parent's wishes.

In addition, the Tenth House and Capricorn are the archetypes through which we all come face to face with our own mortality, and with time and space itself. These archetypes produce the awareness that we have a certain time span in which to fulfill our destinies. Consequently, Tenth House Pluto individuals feel this awareness more acutely than those who do not have Pluto in this house or sign.

This awareness enforces lessons of social and emotional maturity; to grow up and get on with it. It must occur to enforce the awareness of timing in our lives, and that the structural organization of reality at whatever level or application changes or evolves over and through time. The rise and fall of nations, seasonal change, and the evolving nature and structure of an individual's reality all reflect this fact. This awareness, with the necessary reflection, allows for structural change. This change will occur with respect to that which is outmoded or crystalized so that growth at any level of reality can occur. If negatively expressed in an individual, this awareness produces a sense of futility and pessimism: "What's the point?". Positively, it produces the motivations required to accomplish what the individual is here to do.

As with the Fourth House, Tenth House Pluto individuals, in almost all cases, have switched gender in their most recent prior lives; sometimes the current life is the first experience in the opposite gender in many lifetimes. This usually signals that a state of imblance has been reached and further evolutionary growth could not occur through the gender that came before. Thus, for evolutionary purposes, they have switched genders to promote balance and further growth. The resulting hormonal shift of the gender switch can create a variety of emotional moods and feelings that are difficult for the individual to relate to, let alone control. These emotions seem to have a life of their own. The evolutionary need is to become aware of the source or trigger producing the moods and emotions rather than being a victim of them. In this way self-knowledge will grow — what are these feelings telling me about myself?

An example of an individual who refused to accept this gender switch is Christine Jorgensen. Jorgensen was born a man, yet had a sex change operation to become the woman she once was. Jorgensen has Pluto in Cancer in the Tenth House conjunct the North Node in the Eleventh House. The South Node is in Capricorn in the Fifth House, conjunct a Capricorn Moon in the Fourth House. Neptune in Leo inconjuncts the South Node and Moon, and Mars in Pisces is sextile the South Node and trine the North Node. In addition, Venus in Aries in the Eighth House squares the Nodal axis, Pluto and the Moon. Her determination to have this gender switched reversed served as an inspiration and example for others who feel the same need. These astrological symbols suggest confusion within her Soul as to what gender through which to express itself. The square from Venus to the Nodal axis, Pluto and the Moon correlates to "skipped steps."

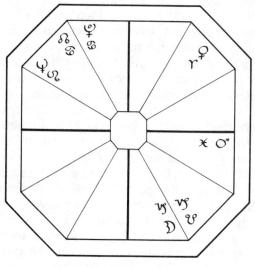

Christine Jorgenson

Pluto on the North Node correlates to the need to continue in the evolutionary direction that have been worked on before, and the Moon on the South Node correlates to the need to relive prior-life conditions that were not resolved before with respect to its opposition to Pluto and the square to Venus. Thus, this individual was born a man who spent the first part of life in that gender. The Soul needed to relate to itself and others (Venus) in that way. Jorgensen was even in the military. This reflects the need to relive conditions from the past that were not resolved in the masculine gender, and to recover the skipped steps in that way. Later on, he sought the sex change operation in order to recover the skipped steps and continue on with evolutionary directions that had been started in the female gender. The Soul now needed to relate to itself and others (Venus) in this way.

The polarity point is the Fourth House or Cancer. The evolutionary intent is that the individual learn to develop internal security, identity completion and a personal fulfillment that is not bound up or linked to the need for success or social position. These individuals are learning how to see themselves without the trappings and security that their social function or career provides.

Beyond this, they are learning emotional lessons. One avenue through which these emotional lessons will occur is in family situations. This can involve either one or both parents, the marriage partner or children or both. The Tenth House Pluto individuals commonly experience emotional shocks delivered to them by one or both parents, their children or a spouse. These shocks force deep self-examination. The loss of a parent can produce this effect, for example. Judgemental parents who do not understand the child's individuality can force the person in upon him or herself. The karmic reason for these conditions in the individual's life is to force self-examination on an inner emotional basis. This process must occur so that they can understand the nature and basis of their emotions and to create a sense of internal security that is not linked to external conditions. In addition, it requires that the individual discover the inner dynamics that create this karmic necessity. If the individual can accept the responsibility for these life conditions, self-knowledge will result.

The family difficulties that these individuals may experience through their own children can force them to examine themselves as never before. They may have "problem" children, or very aware children who produce internal confrontations that force the individual to look at what they do and why they do it.

Under other evolutionary and karmic circumstances, the Tenth House Pluto individual as a parent may neglect his or her children because of career obligations and may never really get to know them. Consequently, the children never really know the parent. The resulting gulf produces guilt that again forces self-examination. In all cases, such emotional shocks and confrontations compel these individuals to examine who and what they are and to examine the reasons and dynamics that have determined their emotional, intellectual, spiritual or physical behavior. They must reflect on how they have identified themselves and what constitutes their security on an emotional level.

Often, the "right" career or social position for the Tenth House Pluto individual is denied or karmically blocked in this life to enforce these same lessons. In other cases, this denial will occur because of prior abuses of power. Being held down in this way enforces the same lessons. Or the individual may be born into a special class or inherited social position. Others, because of prior-life efforts, will succeed in obtaining a position of relative prominence early on in this life. Yet many will lose the position in some way or fall from grace. Some simply will tire of their social position and break away from it. A classic example of a 'fall from grace' is Richard Nixon. His fall from power and the attendant disgrace forced Nixon to examine himself as he never had before.

On a national level it is interesting to remember that Pluto is in Capricorn in the U.S. Chart. The day will come when the United States will "fall from grace." One day it will not be the world power that it is today; it will be one nation among many. This will force a restructuring of the national Soul so that a new collective self-image can evolve. The national sense of security linked with being a world power will have to change as a result of this evolutionary necessity.

An example of someone who simply tired of a social position and made a break with it is John Lennon. Lennon had a phenomenal career that transformed popular culture and touched the lives of countless individuals, yet later in life he withdrew from the scene he had helped create and become a "house husband." When Lennon was a child his father split the scene and his mother left him in the care of an aunt. Difficult family situations occurred through his first wife and son, then through Yoko Ono and the emotional manipulation and control she exercised over him. These experiences caused Lennon to examine himself in deep and essential ways. Raising his son for five years taught him many necessary emotional lessons. Then, after reemerging to pursue his musical career, he was killed by a jealous, deranged, and idol-happy madman. Lennon had Pluto in the Tenth House in Leo, North Node in Libra in the Twelfth House conjunct Mars, which was the ruler of the South Node in Aries in the Sixth House. The ruler of the North Node, Venus, was in Virgo in the Eleventh House inconjunct the South Node. His social and individual karma, from prior and current life actions, had to be fulfilled. What was it in Lennon that led to his assassination in this life?

The evolutionary intent fosters examination in order to discover how identification with career or social position has created an emotional security base. These individuals will necessarily reflect upon their internal dynamics in such a way as to become aware of how they determine their reality, self-image and security. This reflection provides the opportunity to change outmoded or crystallized patterns in identity association. In this way, these individuals will learn how to reorganize themselves and create a new self-image that is not dependent upon career or social function. They will learn to accept responsibility for their own actions and the conditions of their lives and reality. In this way they can progressively learn to internalize their security needs and eventually develop an attitude that even if everything is taken from them, they will still be okay. The more these lessons are realized, the faster potential blocks to a meaningful career will be removed. Their need for social recognition and power can then be met effortlessly through whatever social function their evolutionary and karmic capacity requires.

As these lessons are put in motion, these individuals can achieve a balance wherein their "down" or private time meshes with their external duties and responsibilities. They will learn that the values and beliefs that condition their ideas about what life is and how it should be lived are purely subjective; their judgments apply only to themselves in relation to their own standards of conduct. Consequently, they will learn not to judge negatively the values and beliefs of others. They will learn to accept responsibility of their own lives, and the duties and obligations that accompany them. In so doing, they can motivate and inspire others in the same way. Through self-determination they will be able to realize their goals and ambitions in honest and non-manipulative ways and encourage others to do the same. If they have children, they will learn to become responsible parents who are intensely productive and giving, even if the child is a "problem" child.

Common characteristics of individuals with Pluto in the Tenth House or Capricorn include: cycles of emotional withdrawal, need for social recognition and power, good organizers unless interfered with by other factors, natural leaders, deep and penetrating understanding of how "systems" work, ambitious, serious, pragmatic, anxiety-prone, given to cycles of depression, autocratic and hypocritical.

Famous people with Pluto in the Tenth House or Capricorn:
Paramahansa Yogananda
Ernest Hemingway
Jawaharlal Nehru
Albert Camus
Princess Anne of England

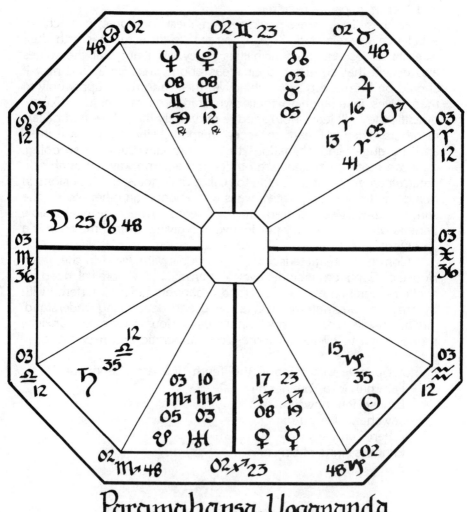

Paramahansa Yogananda
private source

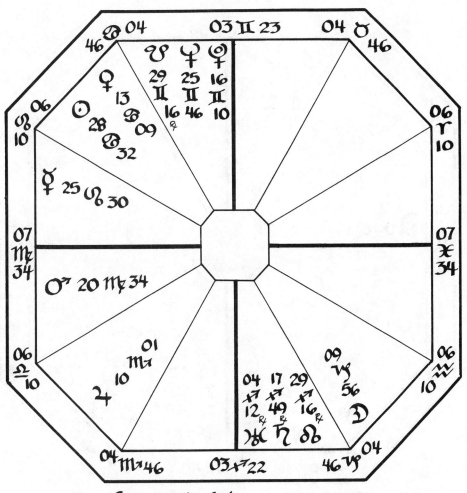

Ernest Hemingway
source: Lois Rodden

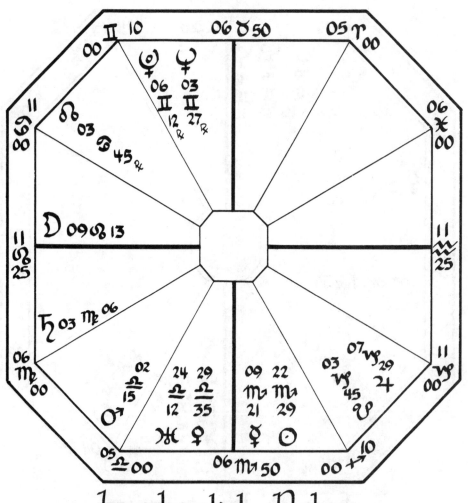

Jawaharlal Nehru
Source: Marc Penfield

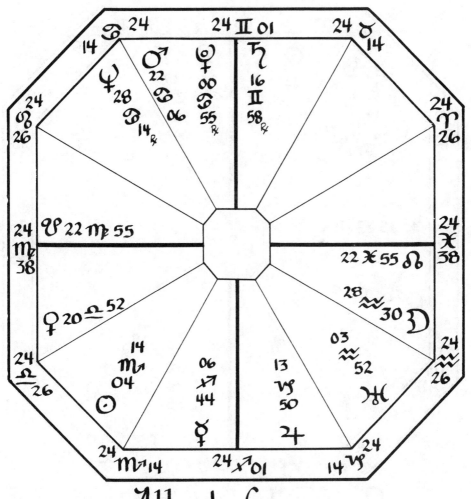

Albert Camus
Source: Lois Rodden

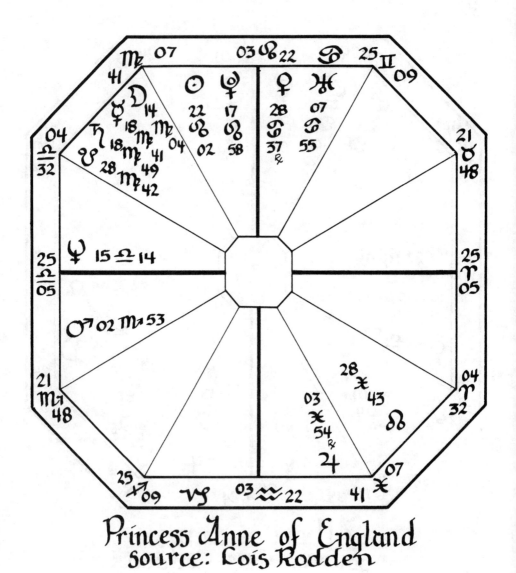

Princess Anne of England
Source: Lois Rodden

PLUTO IN THE ELEVENTH HOUSE OR AQUARIUS

Individuals who have Pluto in the Eleventh House or Aquarius have been learning to break free from crystallized and outmoded forms of self-definition as reflected through the composite effect of society, parents, friends or anything else by which the individual was influenced in his or her early years. In effect, the evolutionary intent has been to shed the skins of the past.

In general, there are three ways in which Eleventh House Pluto individuals respond to this evolutionary intent. Some of these individuals have been learning how to leave behind everything that constitutes the past in this and other lifetimes. In trying to liberate themselves from the shackles of the past, they have learned who they essentially are beyond the cumulative influence of cultural associations, parental input or anything that they have been taught to be or have been led to believe in.

As a result, these individuals have needed to reject any situation that arbitrarily defines them; that tells them who and what to be and how to act. By rebelling they have learned to deflect any external agent that has tried to limit or define them in any way that did not naturally resonate to what they felt they were or were trying to become. In this way, Eleventh House Pluto individuals have been learning how to remove themselves, to pull away from the mainstream of society. They have been learning to resist becoming shackled, conditioned or defined by the customs of society. In addition, they have been learning to view themselves outside the confines of time and space and thereby have developed a sense of objectivity and detachment. It is as if they have learned to stand on the sidelines of human activity, viewing themselves and others from afar. Because of this, they are disconnected from the bondage of any forms of national or cultural identity.

As a result of this evolutionary response, these individuals have also learned to sever all attachments that prevent further growth or the freedom to explore new ways of discovering themselves. These individuals have developed the ability to pinpoint those elements within themselves that are no longer of any use to them. These elements can be emotional patterns, desires, ways of interpreting inner and outer reality and ways of relating to themselves and to society. By focusing upon these outmoded elements these individuals come to understand how each aspect of their personal identity influences and conditions the manifestation of every other aspect, and so the entire structure.

To give a simple analogy, this process is like making a pie. We

199

can entirely change the flavor of the pie by changing or eliminating some of the ingredients. In a similar way, these individuals have learned to observe themselves objectively in order to understand how and why they are the way they are and to change or eliminate any "ingredient" or component that is preventing further growth. Any area of life to which these individuals apply themselves will be closely examined in order to expose that which needs to be changed or eliminated. The actual areas of life to which these individuals apply themselves will be determined by their evolutionary and karmic requirements.

These individuals have a natural innovative capacity because of their inner need to explore new approaches that allow for ongoing personal discovery. It is not uncommon for others to reject or feel threatened by their innovative approach because innovation threatens the traditional ways of doing things. When tradition is threatened, the sense of security is also threatened. In many cases, these innovations are ahead of their time and cannot be assimilated into the existing scheme of things. The change is considered too radical, too much of a departure from tradition. Even the very lifestyles of some Eleventh House Pluto individuals will threaten the mainstream of society, the mainstream of collective beliefs, laws and customs. Many who have needed to experiment through rebellion or rejection of the past, will create new beliefs, customs, taboos, and norms. These are the individuals who have and will be classified by society as bizarre, radical, revolutionary, or just "different."

Some of these individuals will borrow from or align themselves with traditions from one culture and synthesize them with traditions from another culture. In so doing, they will apply them in a new way relative to their existing social/political context or environment. These are the people who live more or less on the fringes of society, seeking out friendships with others of like mind. Some will even attempt to create or establish new social groups within the existing society or culture, or in some way become involved with an established group of people that is also operating on the fringes of society.

In extreme cases, some Eleventh House Pluto individuals in this group will "drop out" of the system or society altogether. They can become anti-social because they feel there is no meaningful role for them to play within society, or no way to integrate or establish their personal and social vision within the existing system. They are called radicals, hippies, outlaws or subversives. In a few cases, some will turn vindictive and attempt to destroy or bring down the system in some way.

The evolutionary necessity to learn lessons of objectivity, detach ment, and the need to sever all attachments that prevent growth can occur through the experience of intense emotional, intellectual, spiritual, or physical shocks. These blows can occur through close personal friendships, casual acquaintances, random meetings with others, or through the mainstream of society itself. Because Pluto in the Eleventh House or Aquarius can create very rigid and fixed ideas about how things should be, how the individual should be, how others should be, or how society should be, others who perceive these ideas as too radical, limited, or unrealistic will deliver the necessary challenges and confrontations. The effect is to induce an objective awareness within the individual relative to his or her attachment to these ideas. The attachment is what prevents further growth. The necessary confrontations can lead to the adjustment or elimination of these ideas so that new patterns can develop.

Yet, because of the compulsive nature of Pluto and the need for emotional security linked with these ideas, the degree of resistance to the necessary changes can be quite profound and intense. If this resistance is maintained, the individual will be isolated by others. As the isolation intensifies, the individual will be forced to examine him or herself, and why others will not deal with them anymore. Isolation can eventually lead to the objective awareness of what the individual is attached to, and the reasons for it.

The Eleventh House and Aquarian archetypes are very important for all of us because they offer the opportunity to revolutionize and transform ourselves; to leave the past behind in order to develop new ideas or ways of relating to ourselves. It allows us to experiment with new thoughts or desires that radically transform who we have been, and who we are now. Through these archetypes we form relationships to others of like mind, those who reflect our own orientation to life and society. In this way, we will not feel too weird because there are at least a few others who are pointed in the same direction as we are.

The Eleventh House and Aquarian archetypes, then, allow us to understand or experience that which is different and unique about ourselves. We experience this "differentness" through individual contrast or comparison, and by forming social bonds with others of like mind, which leads to the formation of subgroups within society. The experiential awareness of how we are different is thus based on observing differences with other subgroups that we do not feel a sense of kinship with. This process occurs naturally to not only promote this awareness, but to also create an awareness in all of us of our connection or relationship to everyone else in an objective way. In this way, we objectively become aware of the individual and unique role that we are playing within our group, within our society, and within civilization as a whole. We become aware of how one part of the whole can influence or change the dynamics of the whole. This dynamic can be applied individually as well as to groups. Social change occurs when enough people of like mind bond together to enforce change within, and relative to, the outmoded structures of society.

Even though these archetypes allow for growth opportunities, individually and collectively, many individuals fear change and fear being or feeling different. Other individuals with Pluto in the Eleventh House or Aquarius have reacted to the prior evolutionary by rejecting or rebelling against the need to be different. They will repress the inner impulse to throw off the past. The feeling of disconnectedness that this impulse produces promotes this reaction. Remember that Saturn, as well as Uranus, rules Aquarius.

All Eleventh House Pluto individuals will come into this life with a deep inner sense of being different from most people around them. This is because of the prior evolutionary intent to throw off the past, to disengage from it. This detachment is projected or linked to the external environment. For those who have rejected or rebelled against this inner need and feeling, the overwhelming need and compulsive desire to belong to a group of people will motivate them to the ends of the earth. In effect, this desire acts as emotional and psychological compensation to block or thwart the inner feeling of differentness and disengagement. These people attempt to align themselves with the prevailing social norms and laws of society. As long as everyone else is doing it, whatever "it" is, they will do it too in order to belong to the group. Any social group creates "norms" of behavior, and the overt or covert pressure to conform to those norms. Individuals reacting in this way have become compulsively and utterly dependent upon their peer group for a sense of personal fulfillment and meaning. Their individuality, their purpose, their values and beliefs are simple extensions of the "norms" of their peer group. Nevertheless, despite this compensation, these individuals still feel different. It is this feeling that to some extent promotes the necessary objectivity and detachment. No matter what the individual does in order to feel "normal", to belong, they will still feel different. This feeling will promote the typical Plutonian question "why?" Relative to the free will that we all have, these individuals have been learning to make choices of whether to pursue that which is different in themselves regardless of where it may lead, or to deny or suppress that which is different regardless of the consequences. Some have taken a few tentative steps to break free, only to retract and try again.

Others have taken quite a few steps and are finally under way towards pursuing and developing their socialized individuality in some way. Others remain completely locked into peer group control.

The third segment of individuals with Pluto in the Eleventh House or Aquarius have reacted to the prior evolutionary impulse by rabidly defending the traditions of the past. These individuals have feared being different; have feared experimentation, new visions, new models, and new ways of doing things. Instead of disengaging and detaching from the past, these people have manifested a fixated death grip upon the past — a totally Saturnian reaction. The first segement of people manifested a Uranian response to this prior evolutionary intent. The second segement of people manifested both Uranian and Saturnian responses. This last group is Saturnian. Yet, even people who respond in this way will feel different because they will sense that "time" is passing them by. These people are social dinosaurs. Individual and collective life is always growing, always in a state of becoming, of evolving. It is exactly this obvious process that these people fear because almost any change, any departure from tradition, is perceived as undermining or threatening the old order and stability of the way it used to be.

These individuals use the old order as a panacea to correct the perceived ills of the moment, and to project into the future "the way" that it should be. Rather than allowing themselves to develop a new vision to apply to a changing world, they project an old, tired vision upon the future as the cure for the problems of today. This orientation reflects the fact that these individuals are inwardly suppressing, in a desparate way, the inner impulses to "shed skins" from their own prior lives and the patterns of identity association that are based on them. They too will form social bonds or friendships with others of like mind. As individuals, and as a group, they can develop objectivity and detachment through the isolation of their subculture within the context of society. These individuals, and the subgroup, are different when compared to others who do not respond to their sociological ideas and identities.

Differing or interacting combinations of these three distinct ways of responding to the Eleventh House or Aquarius archetypes give rise to all the possible subcultures within a society. In a negative expression of the Eleventh House or Aquarius archetype, one group or a combination of groups may attempt to influence, control, or eliminate another group for reasons of security and power. This effort reflects the egocentric desire for separateness and can promote narrow-mindedness and attachment rather than objectivity and detachment. An individual can react to another person in the same way for the same reasons. Positively expressed, group and individual differences will be tolerated, accepted and encouraged. Combinations of positive and negative reactions occur when the extreme fringe groups, whether ultra-liberal or ultra-conservative, are perceived to threaten the stability or security of the entire group.

Clues found in the birth chart indicate which way an Eleventh House Pluto individual is likely to gravitate regarding group identification. If the weight of the whole chart suggests conformity and emphasis upon tradition, the individual will probably identify with the mainstream of society. If the chart suggests nonconformity, the individual will likely identify with fringe groups. If the chart suggests heavy introversion, the individual will have few friends, and no particular group association; he or she may possibly be a group of one.

In addition, we also need to consider the evolutionary condition of the individual. Those in the herd state will belong to mainstream groups. Those in the individuated state will belong to more independent minded groups. Spiritual state individuals will associate with spiritual groups. Linking evolutionary condition to the energy balance of the chart, i.e. introverted, conformity and so forth, will give a fairly good understanding of which subgroups within a society the individual will feel a kinship with. Some examples to illustrate this point would be: an introverted/ spiritual person would probably withdraw from any group interaction; an introverted/conformity type would be a silent follower of the mainstream peer group; an extroverted/individuated type may be a leader for a splinter group, and so forth.

Pluto's sign will give us additional clues. For example, a person with Pluto in Virgo may feel intensely shy and intimidated by social groups. He or she may feel critical of the group they are connected with and the group may be critical of him or her. This individual may succeed in finding another group that reflects his or her own ideas, and from the bunker of that group may hurl critical hand grenades at all other groups that are different. As an example, many people who belong to the punk rock group prevalent in society today have Pluto in Virgo. As a group, and as individuals, they tend to react in the way described.

Let's use a case history to illustrate these dynamics in detail. This woman has Pluto in Cancer in the Eleventh House. The South Node is in Gemini in the Tenth House. The planetary ruler, Mercury, is in Aquarius in the Sixth House. She is in the individuated evolutionary condition. She was born into a black lower-class family in Los Angeles, California. Her parents provided the best they could for her, and her father worked as hard as he could to save money for her education. He refused to allow her to get stuck or victimized by her race or subcultural circumstances. His message, early on, was that she could become what she wanted if

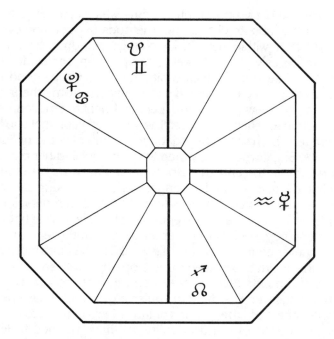

she made the effort and was determined. Her immediate environmental circumstances were to witness the poverty and suffering of her neighbors, and the "messages" from society about being black. When she was around fifteen she decided to become a social and community worker so that she could help improve conditions for her race. She refused to become trapped or to give in to the racial discrimination of her times. She refused to wallow in self-pity and resolved to do whatever it took to get herself through school. She knew deep within herself that she was different, yet she totally understood the reasons why her people felt as they did. She identified with and understood the social conditions that promoted this kind of consciousness so well that she resolved to change them through the work that she had decided to pursue.

She did get herself through school with the assistance of the money her father had saved, and through part-time work. She earned a Ph.D. in social psychology and returned to her community to implement work training programs. She helped pioneer block grant programs to repair the community, providing work for people in the process. She initiated a wide variety of workshops and seminars for the community on many subjects. She helped revolutionize her community. Her actions and programs became the model for programs in other areas.

One of the major problems that she encountered was based on her need to impose her vision on others. She was extremely defensive of criticism, and unable to change her methods of operation even when she knew better within herself. To change or admit error was sub-consciously perceived as a threat to her position, social identification and emotional security. Thus, despite the tremendous good she was doing, she attracted criticism and confrontations from others. Another problem was that her whole sense of personal identity, emotional security, and orientation to reality was based the social position that she held. As time changed she had trouble adapting to the changing needs of her community. She compulsively tried to keep applying her old ideas on new situations. In the end she experienced a nervous breakdown that incapacitated her. She could no longer work even though she was still a relatively young women. She was forced to deal with herself in a new way. As time went on, she healed enough to be able to write books about social psychology, social justice, and novels with these themes.

In whatever condition we find Eleventh House Pluto individuals, all have had the evolutionary need to develop objectivity and detachment, to sever attachments that are preventing further growth, and to form group bondings with like-minded souls. In general, all have needed to deflect the impact of other groups or individuals in order to deepen the awareness of their own individuality. All have needed to link their individual purpose to a socialized purpose, a group purpose, and a socialized reason for being. Depending on evolutionary development and karmic dynamics, some Eleventh House Pluto individuals have had, and will have, the capacity to lead a group of people or an entire nation. Pluto's natural magnetism can attract others who perceive that the Eleventh House Pluto person had and has something powerful to offer.

Many of these people, based on prior development, possess an inherent ability to understand group needs, and to uncover the basics of group dynamics in order to understand all the individual roles and functions that constitute a group. Those who have leadership capacity and are "future oriented" can develop comprehensive social visions as to how a group of people or a society should change in order to progress and grow. They will experiment, try new ideas, and adjust as necessary. In certain cases, they will experiment with new ideas but then resist changing the blueprint even in the face of intense confrontation, thus bringing about their own downfall, or the downfall of their ideas. Others may attempt to affix old visions upon new situations. Such a leader may attract followers for a time, but because the ideas are old, a downfall is usually guaranteed. On another level, Pluto's natural repulsion can also create total isolation for some individuals with Pluto in the Eleventh House by an entire group or subgroup. We must look to the total nature of the individual's karmic signature and background in order to determine the evolutionary and karmic necessity for this situation.

The polarity point is the Fifth House or Leo. The evolutionary intent is one wherein the individual must learn how to take charge of his or her own destiny. They must learn how to implement their ideas about themselves, that which they see as possible for their lives, the goals they want to accomplish, and the ideas pertaining to future possibilities. One of the most common symptoms that individuals with Pluto in the Eleventh House experience is thinking about the possibilities pertaining to their future which go unacted upon. This symptom is a reflex reaction to the prior evolutionary intent to break free from the past. In this life, this effect only creates a dissatisfaction with the present as long as the ideas are not acted upon. Thus, the current evolutionary intent is teaching these people to translate these ideas into action through taking charge of their own destiny and shaping it along the lines that these thoughts are suggesting to them. In order to accomplish this, they must learn to minimize their security needs and patterns of depending on others to tell them that it is okay to do whatever it is that they want to do. They must also learn not to wait around until everyone else acts first, and not to wait for others to support their ideas regarding strategies for themselves and others.

Those who have been repelled by others or society, who have been forced to stand on the sidelines, must learn to develop their own individual purpose and link it to a socially relevant need. They cannot remain standing on the sidelines, hurling stones and insults at a system that is not behaving according to their desires.

In general, these individuals must learn that their power lies in the fact that they *are* different, and that it is *okay* to be different. By learning how to act and implement, they can become effective leaders in any area in which they are evolutionarily and karmically destined to play. All individuals with Pluto in the Eleventh House have the inherent power to act as instruments of innovative and crea-tive change in whatever area of life that they choose. Even the rabid defenders of the past can exert a positive influence as stabilizing agents if changes are happening too quickly or are obviously off course.

All of these individuals must learn how to detach, objectify, and sever their most cherished visions as to how things should be, when those ideas or visions are not appropriate to the situation at hand. Through environmental challenges or confrontations they must learn to objectively and impersonally change what must be changed. The potential trap is to remain detached in stoic defiance against those who do not agree with them — not to change.

All of these individuals can experience emotional shocks, disappointments, leavings and rejections through friendships. Having the rug pulled out from under their feet in this way occurs to enforce the lesson of taking charge of their own destinies. The severing of dependencies on others cannot help but to enforce this lesson. In addition, this kind of event will enforce the lesson as to what really constitutes a friend and what does not. In general, these people will only have, or be allowed, a few close friends at any one time in this life. In certain cycles, they may experience a total lack of friends.

By learning these lessons, Eleventh House Pluto individuals can play a variety of socially meaningful roles. Their natural inventiveness and creativity can shine. The resulting metamorphosis will produce self-confident individuals who possess the power to understand objectively who and what they are, why they are the way that they are, and how best to actualize their own creative purpose within the context of a social need. In this understanding, these individuals can assume socially relevant and meaningful roles, roles that have the power and potential to transform the existing barriers that are restricting further growth and evolution in the area of life to which they have applied themselves. In the same way, these individuals can promote this understanding in other people, even in whole nations.

Common characteristics of those with Pluto in the Eleventh House or Aquarius include: behavior ranging from being extremely anti-social to following the crowd to rabidly defending tradition, intrinsic feeling of being different, obsessive and compulsive thought patterns, innovative, unique, creative, a good friend, cycles of utter detachment within cycles of intense focus upon themselves, potential for sudden and erratic behavior, iconoclastic, hard to really know or define correctly, aloof.

Famous people with Pluto in the Eleventh House or Aquarius:
    Albert Einstein
    Jesse James
    David Bowie
    Benjamin Franklin
    Lyndon B. Johnson

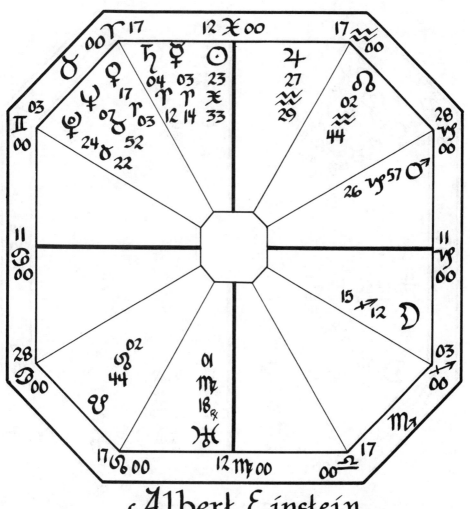

Albert Einstein
Source: Marc Penfield

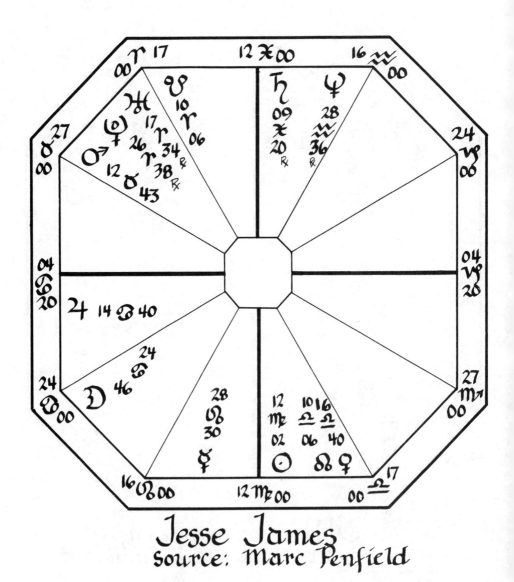

Jesse James
Source: Marc Penfield

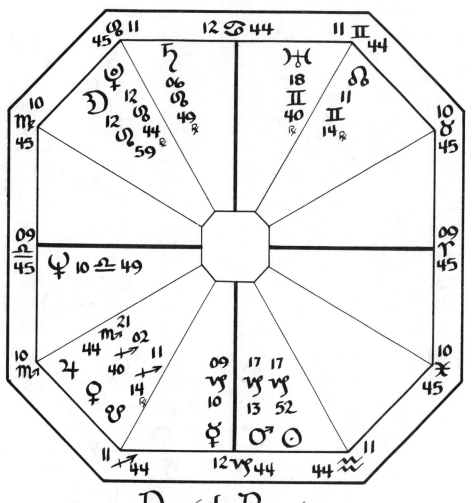

David Bowie
private Source

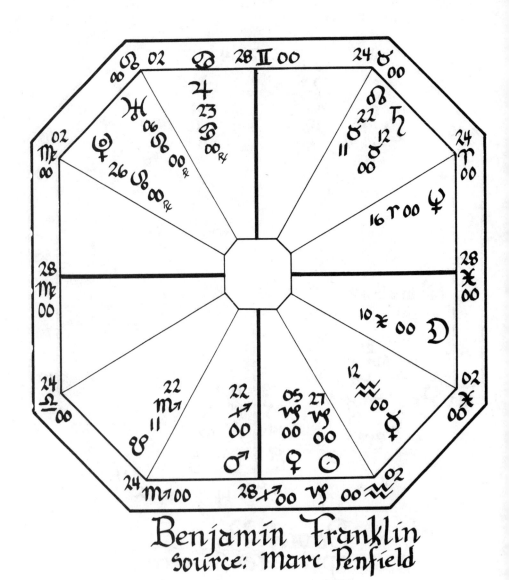

Benjamin Franklin
Source: Marc Penfield

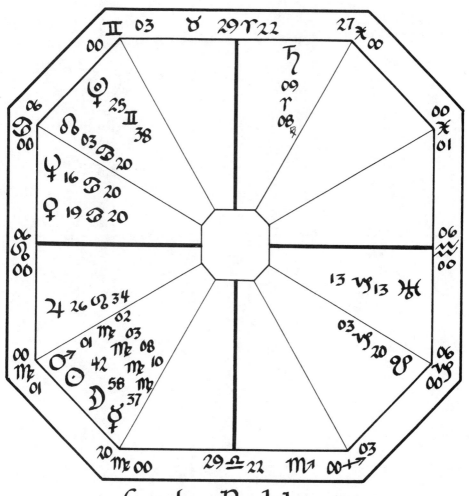

Lyndon B. Johnson
source: Marc Penfield

PLUTO IN THE TWELFTH HOUSE OR PISCES

Individuals who have Pluto in the Twelfth House or Pisces have had the prior evolutionary intent or desire to align themselves with some type of transcendental belief system in order to realize the unity of all of Creation, and to see or experience their own individuality as an extension of the Source of Creation. In so doing they can spiritualize all aspects of their lives.

In general, the Twelfth House and Pisces are the composite of all other houses, signs and planets. It is the archetype that describes the totality of life and reality in this and other planes of existence. Simultaneously, it is also the archetype in which the Source of all that constitutes life and reality is sensed and must be consciously developed and realized by the individual.

Those with a Twelfth House Pluto have had an emphasized and direct desire to dissolve all barriers preventing the merging of their individual power into the cosmic whole in order to experience or realize the Ultimate Source of power. These barriers can be emotional, intellectual, physical or spiritual in nature. The evolutionary need to dissolve all the old barriers preventing direct individual identification with the Source has demanded or required that these individuals align themselves with a transcendental belief system through which the spiritualizing process could or can take place.

There has been a desire and need to seek out and realize values, beliefs or knowledge that transcend time and space, and that transcend culturally based values. Whereas the Tenth house or Capricorn archetype produced the realization of the individual's own mortality from an egocentric point of view, the Twelfth House or Pisces Pluto produces the knowledge or realization of the individual's immortality, of infinity from the Soul's point of view. The Tenth House or Capricorn correlates to time and space. Pisces and the Twelfth House correlates to timelessness and infinity. The process of expansion that essentially started in the Sixth House and Virgo culminates in the Twelfth House and Pisces through a metamorphosis that produces a merging of the individualized ego with the Source.

As a result, Twelfth House Pluto individuals will come into this life with a deep inner sense that they are standing upon a precipice. Behind them is the light of the known world (their past, and that which symbolizes culture, time, and space); and in front of them lies the darkness of infinity; the abyss (their future, and that which symbolizes timelessness and the Universe). Standing upon the precipice, these individuals face choices as to which direction to go: backward, forward, or to remain upon the precipice itself, paralyzed from mov ing in either direction.

The choices that the individual has made in the past will influence what is experienced in this life relative to this evolutionary desire, intent, and necessity. These choices will determine in what state or condition we find the individual in this life.

We need to find out how the individual has responded to this prior evolutionary intent. As in all houses and signs, the coexisting desires in Pluto and the Soul, one for return to the Source, one to separate from the Source, interact to determine the individual's respon-ses and drives. Yet, with Pluto in the Twelfth House, the very nature of the prior evolutionary need was to merge with or identify with the Source of Creation; to universalize or spiritualize.

Thus, the very essence of egocentric identification within the individual was being eliminated and dissolved. Any form or applica-tion of separation from the Source has been subjected to this prior evolutionary pull to dissolve the separating desires. The three primary reactions to this Twelfth House evolutionary process will describe choices that led to actions, and thus to the evolutionary and karmic conditions that these people will experience in this lifetime.

Before we detail these three reactions and their consequences, let's discuss the most common inner and outer experiences that these individuals will face as a result of the prior evolutionary intent. In general, most of these individuals have effected some degree of resistance to the evolutionary pull to merge, surrender and seek identification with the Source. The resistance triggers cycles of con-fusion, disorientation, alienation and disintegration in varying degrees of intensity throughout their lives.

Separation, in whatever form it manifests, breeds reisistance and repulsion to the merging and surrendering process. Separating type desires imply that most of these individuals have an unconscious fear of losing control of their lives in some way. Consequently, most Twelfth House Pluto individuals can desperately latch onto anything of an individualizing nature in order to feel in control of their lives. Yet, because of the evolutionary pull, they also experience the inner sense that whatever they have individualized and given personal identification to, is not really them, they can't really relate to it. There is something more; a missing link that they can not put their finger on. On a cyclic basis, this inner sense intensifies and becomes more dominant in their consciousness. As this occurs, confusion, aliena-tion and disassociation take place. Deep inside these individuals will inherently feel diffused and unintegrated during these cycles. As a result, many will overly identify with a single element or aspect of themselves, and will manifest everything else through this one aspect in order to induce a sense of individuality against the backdrop of inner diffuseness. In some cases, these individuals will identify with or assume the beliefs, causes or identity of a subgroup within the collective whole in order to create this sense of egocentric individuality. This area or aspect of identification is given tremendous power because it serves as the way and vehicle to relate to themselves as an egocentric individual.

In certain cases, these individuals have been, or can become, powerful icons because of the power expressed or projected through that one aspect or dimension of themselves. Some have or will exhibit a larger-than-life aura or impact on others Through the power of their belief in who they are, and what they are doing, they have the ability to captivate the interest, imagination, support or persecution of mass consciousness. The life of Jane Fonda reflects these dynamics. Fonda has Pluto in Cancer, retrograde in the Twelfth House. The South Node is in Gemini in the Tenth House, its planetary ruler, Mercury, is retrograde in Capricorn in the Sixth House. Jupiter, ruler of the Fourth House Sagittarius North Node, is in Aquarius in the Sixth House opposed to Pluto. In the earlier part of her life, Fonda identified with the anti-war movement during the Vietnam era. By focusing herself upon this social identification and movement, she became

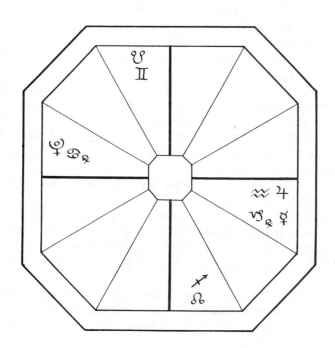

a spokesperson for it. Her journey to North Vietnam during the height of the war brought her social persecution or praise depending on which side of the issue each individual identified with.

Overly identifying with a single aspect of the self induces compulsion because at least that one aspect is familiar and known. To change this way of relating to themselves threatens to induce the unknown. Consequently, it threatens a loss of egocentric control and, therefore, personal insecurity. Yet, the evolutionary impulse to dissolve such personal limitations will necessarily create confusion, disbelief, disassociation and alienation from the one aspect of themselves that these individuals have most identified with. During key cycles, Twelfth House Pluto individuals will experience as essential implosion leading to a breakdown or disintegration of some aspect of their identity or beliefs. As the implosion process intensifies, they experience an essential confusion and disorientation. The realizaiton dawns: "Is this really me?" — or "that's not really me" or "there's more than this." As the process unfolds, these individuals experience a progressive sense of being out of control or losing control of their life. Furthermore, nothing immediately replaces that which is being dissolved. Feeling out of control, lost and confused they are forced to experience cycles in which nothing means anything, and they don't have the power to make their life mean anything. Some will thrash and convulse in a compulsive and panicky attempt to resurrect an old meaningful pattern of relating to themselves and to life in an attempt to return to the light of the known world; the past.

These reactive compulsions are not only desperate, but doomed to failure. They may work for a time, but in the end will dissolve into meaninglessness. Wandering in confusion, many of these individuals will magically experience a metamorphosis in which a new thought, realization, dimension of themselves, or a new way of relating to themselves and life emerges from the Soul. This process seems to occur of itself, and is not a product of the individual's egocentric deductive mental process.

These new patterns and thoughts that emerge out of the Soul are actually "inspired" by divine sources, although many of these people would not consciously acknowledge or identify it as such. The new situation is simply born of itself. Now clarity exists where there was confusion, integration where there was disintegration, belief where there was disbelief, relatedness where there was alienation.

This process occurs to teach these individuals not only belief, but also the awareness that they are connected to a much larger living whole; the universal. They have been learning how to experience themselves as an individualized wave upon the sea. This process is, and has been, teaching them about the areas or dimensions within that are preventing and limiting their personal identification within the cosmic whole; to shift their center of consciousness from the wave to the sea. This process is teaching these individuals, willingly or unwillingly, to plunge into the abyss of infinity, rather than remaining paralyzed upon the precipice, or turning backward to the light of the past.

In plunging into the abyss of infinity, Twelfth House Pluto individuals are learning the lesson of faith; that the fear of individual dissolution, or surrender to a higher power than themselves, is only a delusion of separating desires reflected through the ego. They have been learning faith by willingly or unwillingly plunging into the abyss of the cosmic sea and experiencing, on a cyclic basis, disintegration of whatever personal limitation which needed to be dissolved because it was promoting a non-growth situation. Coming out of this plunge into the abyss, a metamorphosis will have occurred. This metamorphosis will produce a new realization, the discovery of a latent dynamic within the individual, or a parting of the mists of confusion that reveal the answers and new directions that they were seeking. The cycles of dissolving that lead to cycles of clarity symbolized by the plunges into and out of the abyss, occur of themselves. The individual cannot control them with respect to his or her ego. These cycles occur through the direction of the person's Soul with respect to the intrisic intent of the Twelfth House Pluto "life lessons." In this way, these individuals are learning faith. Faith is linked to the experiential awareness that somehow, or in some way,"something" has made, and will make, the past and current difficulties all right. By experiencing revelations and realizations that seem to arrive spontaneously in their consciousness, these individuals will be experiencing, in their own ways, "divine communion" or guidance.

Many individuals with Twelfth House Plutos have succeeded in repelling or denying the evolutionary impulse to merge or identify with the Source. Through denial, these individuals will create one fantasy or illusion after another in order to find the meaning that they are seeking. It is as if the dream or fantasy symbolizes the Ultimate Meaning. The denial force reflects the desire to separate from the Source. The actual nature of the dreams or fantasies are conditioned by the specific kinds of separating desires that these individuals have. Because all Twelfth House Pluto individuals are seeking a sense of ultimate meaning in their lives, these dreams and fantasies are given tremendous power. They are glamorized and distorted out of proportion. They become very real to these individuals and are focused upon in a singular kind of way. They become potential experiences through which the individual will seek to discover him or herself. These fantasies and dreams can become potential experiences for two reasons:

1. The desires inherent in Pluto translate into a personal will to actualize those desires. In the context of the Twelfth House, this process teaches the individual that he or she is a co-creator of reality in relationship to the Ultimate Source. The power to actualize these dreams and fantasies is thus linked to visualization and belief. If the individual believes in the dream or fantasy enough, the power of visualization and affirmation intrinsic to Pluto in the Twelfth House, it will come true.

2. The fantasies can teach the individual about the nature of his or her dream illusions and delusions. An important point to remember is that these dreams and fantasies can seem very real to these individuals either as they are creating them in their consciousness, or as they are actually being lived out. Even to an outside observer the actualization of these separating type dreams and fantasies that the Twelfth House Pluto individual creates will seem very real. An observer would be hard pressed to see or understand that these "realities" are actually based on illusions and dreams because the individual has succeeded in actualizing them: they are living them out. As such, they constitute the basis of the person's actual or concrete reality. They seem so real.

Let's use a case history to illustrate this point. This individual has Pluto conjunct Mars in Leo in the Twelfth House. The South Node is in Sagittarius in the Fourth House. Its planetary ruler, Jupiter, is in Libra, retrograde in the Second House. The North Node is in the Tenth House in Gemini, and its planetary ruler, Mercury, is in Taurus also in the Tenth House. A Seventh House Pisces Moon squares the Nodal Axis, and is inconjunct Pluto and Mars. This individual is in the individuated evolutionary state. He has a natural awareness of larger forces than himself coming into this life. Yet, he also had strong unconscious fears of being pulled too far into the abyss, of committing to the conscious development of his intrinsic spiritual need and nature. He had an unconscious "god complex" wherein he expected reality to revolve around him. He also subconsciously desired to play god in other peoples lives. He had natural teaching and healing skills coming into this life based on prior-life efforts. As a result, he created new healing methods by synthesizing a variety of related disciplines. Because of his desires and needs to be considered important and powerful by others, to be recognized as special by others, he visualized and created a career that was based on this new healing discipline

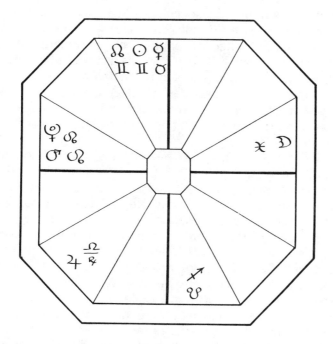

and approach. In addition, he desired to create a nationwide network through which this healing discipline could be applied.

Beyond the desire to be recognized as all powerful, he also had deep desires to make a lot of money through his work. He attracted other people who had similar healing desires as part of their own sense of purpose in life. He fueled their desires through teaching them these new healing methods, and by promising the Moon with respect to the levels of money and fame that they themselves would realize by joining forces with him. By creating this kind of network he would feed all of these people by sending them, on a regional basis, the clients who were seeking his services. In other words, he was the apex of the pyramid forming this national structure. In this way, he also made money from the work of the people that he was setting up to fulfill his dream. He became an icon for others who knew him in this capacity. He did in fact actualize his desire and dream. He tried to integrate himself around this one aspect of himself, through a continual workload that involved writing a book, lecturing, and organizing the national network. The compulsion to work continually was a reflection of his fear of being pulled into the dissolving nature of the abyss. Yet, the pull of the abyss promoted a fracturing of his self-image that was limited to the healing dynamic.

This fracturing or disassociation effect promoted a sense of ultimate aloneness, a sense that something was missing or lacking in his life. Thus, he checked another career dream wherein he desired to write romantic sex novels. He even assumed a pseudonym. Through the plots and characters in his fiction writing he vicariously fulfilled his own subconscious desires. Because this "happened" to coincide with a lack of sexual romanticism in his real life relationship, he became involved in an affair that resembled the plots of the books he was writing.

This individual did in fact materialize his subconscious desires and dreams. They seemed, and were, very "real" to him and others. He had no conscious realization that he was avoiding and/or fearing a merging of himself with the Source. He had no conscious idea that he had the desire to "play god," or that he felt ultimately alone. In fact he felt quite to the contrary, a common problem with most Twelfth House Pluto individuals. They have a difficult time recognizing the actual basis of their "dream realities" because they are living them. They are hard to recognize because the motives and fears that create these desires and realities are quite unconscious. Yet, because these desires are separating by nature, and are not the result of a conscious direction that is based on an inner relationship with the Source, the realities that are created are destined to fail in some way. The dream balloon will pop, and reality will strike. These individuals then confront the fact that this "reality" was only a dream, the fantasy only a fantasy, and that, in the end, they are left standing on the precipice alone, without meaning, contentment or inner peace.

In the above example, the national network that this man created collapsed because many of the workers within the network began to see through him. It collapsed because many of these people were not qualified to do the work, and because he did not exercise proper discrimination or judgment with respect to whom he could heal or help, and whom he could not. In the last analysis, it collapsed because of his wanting to play god, and because this work was not by the direction of the Source. In other words, the need in all Twelfth House Pluto individuals is to align their personal wills and desires with the will and desires of the Source. When this does not occur, then that which is created through separating and egocentric desires will collapse at some point. This emotional shock intends for the individual to realize the nature and basis of his or her dreams, illusions and fantasies. Just as the national network began to collapse, the affair that he manifested based on his sex novels also blew up in his face. The fantasy constructed in his mind did not equal the reality of the actual affair which almost ended his existing relationship. The emotional shocks of reality from both situations set in motion a cycle in which the individual had to totally reevaluate his life; where was he coming from, and for what reasons.

The compulsive chasing of the dream can last many lifetimes because the dreams and fantasies are limited only by the power of the individual's own imagination. At some point in the evolutionary process, however, these people will exhaust their separating and avoidance oriented desires, and an ultimate weariness will descend upon them. Disillusioned, they will finally turn to the Source to face the abyss of infinity and seek a relationship to the Cosmic Whole. In this exhaustion of separating fantasies and desires that issue from a deluded imagination, all Twelfth House Pluto individuals will one day realize the old spiritual axiom that "life is but a dream." As the wave upon the sea must return to its source, the ocean, so too must all of us return to our Source. Once these individuals come to this realization, they will become divinely inspired as to the nature of their real identity, and can also become divinely inspired to fulfill some type of "mission" on behalf of us all. This realization will occur as soon as the individual develops a conscious relationship with the Source so that his or her actions and desires are in direct conformity with how the Source wants to express Itself through them.

Resistance and denial to merging with the Source can be manifested as other "symptoms" including denial that anything is wrong with themselves, and their lives. Compulsively they pretend that everything is OK. Even though their "reality" may not be what it seems, with the blind strength of their Twelfth House Plutonian wills they try and make it OK whether it is or not. At some evolutionary and karmic point, the proverbial rose-colored glasses will break. Cracks in the tint now allow the glaring white light of truth to illuminate their actual reality and their life situation. The cracks are induced through unconscious self-undermining activity, or through environmental confrontations in which the self-created reality undermines what they believed in. The shock of experiencing reality creates the necessary disbelief, disassociation, confusion, and alienation that enforces the awareness as to how the person was "hung up"; of that which was actually unreal. These kinds of shocks enforce the awareness of why the individual was blocking, denying, being excessively naive, seeing what they wanted to see or refusing to acknowledge that anything was amiss. This awareness usually occurs after a necessary amount of time is spent in not knowing why. One day, of itself, the necessary knowledge illuminates the individual's consciousness. When the magic of this clarity and understanding occurs, the individual is forced to consider that some other Power is operating in his or her life.

All Twelfth House Pluto people will come into this life with an ultimate, timeless sense of right and wrong. This ultimate sense of right and wrong is reflected in a standard of idealized conduct. This standard of idealized conduct manifests itself in varying states of conscious development within these individuals, depending on their evolutionary state. In most, it is sensed as a deep inner feeling of how things should or could be. This inner sense has developed because of the prior evolutionary intent to seek identification with a transcendental truth or reality. As a result of this prior evolutionary intent, all these individuals judge themselves, others, and life in general in reference to this idealized standard of conduct. When their own actions or the actions of others, or humanity's actions do not reflect this standard, they will commonly judge themselves, others, or humanity in a negative way as measured against the ideal. Because of the dual desires in the Soul, the desire to return to the Source competes with all the separating desires that are manifested as dreams and fantasies of an escape or avoidance nature with respect to merging with the Source. Thus these individuals will, on a cyclic basis, judge themselves in relation to what they should have done, or should be doing. As a result, many exhibit an intense self-induced persecution because of their perceived inability to do the "right" thing; to conform to the idealized standard of conduct whose ultimate root is based in the evolutionary need to merge with the Source.

Consequently, many will also persecute or judge others relative to their own weaknesses or shortcomings. Many have drawn lifetimes of persecution upon themselves in order to atone for their guilt for not doing the right thing. This condition serves as the basis of pro-jection wherein they can judge and persecute others for not doing the right thing according to the Twelfth House Pluto individual's standard of idealized conduct. Some who have drawn lifetimes of persecution upon themselves will come into this life with an ultimate sense of victimization or martyrdom. They will feel that life is beyond control, or that they are at the mercy of forces beyond their control. Some of these "victims" will compulsively avoid or escape reality through drugs, alcohol, or any avoidance-oriented activity. This prior life karmic condition usually reflects the fact that these individuals have desired to undermine and erode the illusion of power from an egocentric point of view. Many will have abused power from an egocentric point of view before — the "god complex." Thus the necessity for a few lifetimes of persecution and containment wherein their lives were at the mercy of forces apparently beyond their control. Those who have failed to realize the intent for these conditions, to align one's will with a Higher Will, have or will compulsively blame everything and everyone to justify their negative and avoidance-oriented behavior. Some will remain locked in their own self-imposed prisons, quietly wasting away. Still others will be compulsive "Florence Nightingales" who attempt to help everyone, whether they want this help or not. These types will shower indiscriminate mercy upon all; judging nothing, as if judging itself were to be avoided: to be treated as a "wrong." In reality, Pluto Twelfth House individuals have needed to learn what judgment really means, what it is, what it is not, and how to exercise proper judgment upon any aspect of life.

The lesson here has been and is to learn that "ultimate" judgment is based upon intentions. Intentions reflect desires. Thus these individuals must learn to focus upon their desires in order to understand the connection to their intentions. If the desires for right behavior as related to their idealized standard of conduct is consistent and strong enough, the intention to act in that way will follow.

Similarly, Twelfth House Pluto individuals must learn to judge others according to their intentions. The inner judgment, which is based in their own actions, and how this judgment is projected on others in relation to their actions, is the important issue here. As Jesus said when the prostitute stood before the avenging crowd, "Let the one who is without sin cast the first stone." Of course, nobody could throw the stone. The crowd based their anger on a standard of idealized conduct. Yet, as Jesus pointed out, even though all of them had the desire to conform to these standards themselves, none of them had perfectly done so. However it was appropriate to inwardly judge the error in the prostitute's ways, as it was appropriate to judge the errors in their own ways against the standard of right conduct. It was not appropriate to hurl stones at the prostitute because of her error. It was not right because they were not free of error or sin themselves.

The inner judgment was necessary because the judgment itself promotes the vehicle to become more perfect in relation to the standard of right conduct. The key in this necessary judgment is to forgive oneself and others for the shortcomings or failings that occur because very few of us are perfect. The intention and resolve to become perfect, to improve, is the ultimate standard upon which all of us must be judged by ourselves, others, and the Source. It is important to exercise this kind of judgment on ourselves. In the same way, Twelfth House Pluto individuals must utilize an ultimate standard of conduct in order to recognize the incorrect desires and behavior of others so they do not reflect, assume, or manifest the "wrong" behavior of others through lack of necessary judgment and discrimination. Concerning others, these individuals must learn not to express judgments upon others unless asked. Unless asked, the best strategy is to mentally affirm change for them.

Some who have been karmically destined to play the role of a martyr or victim in order to realize their own lessons in judgment have left us with inspiring testimony to this kind of Twelfth House truth concerning judgment and forgiveness. Jesus, of course, is a paramount example — "forgive them Father, for they know not what they do." Anne Frank, another example, wrote in her diary: "In spite of everything, I still believe that people are really good at heart." Jesus had Neptune conjunct the South Node in Scorpio, and six planets in Pisces opposed Pluto and Mars retrograde in Virgo. The ruler of Jesus's North Node in Taurus was Venus in Pisces. Anne Frank's chart has the South Node in Scorpio, ruled by a Twelfth House Pluto, with North Node in Taurus ruled by a Tenth House Venus in Taurus.

The evolutionary lesson in proper judgment is extremely important in another respect. Many Twelfth House Pluto individuals will have an irrational and abject fear of the unknown. Many will not be able to face themselves alone. The Soul for all Twelfth House Plutos is anchored in the universal or cosmic whole. While this is true for all of us, Twelfth House Pluto individuals are pulled into the Universal Source much more intensely than those with any other house posi-tion of Pluto: it is their emphasized and intensified "bottom line." Consequently, many will experience this pull as a deep inner vortex, like a black hole in the universe, that threatens to consume them in the abyss. Fearing dissolution and loss of control, they can manifest all forms of aberrant behavior. Acute paranoia, neurosis, schizophrenia and phobias are extreme manifestations. Less acute "symptoms" are intense nightmares, sleep walking, fear of closing their eyes, can't be alone, always talking, intense escape or avoidance patterns (always has to be doing something until physical exhaustion overcomes them), and refusal or denial of anything they don't want to hear, touch, feel, experience, taste or smell.

Some with Pluto in the twelfth House can be the recipients of "visitations" by other entities or energies. Or, they can experience feelings, emotions, moods, thoughts, or desires of "unknown" origins. Sometimes the origins are inner repressions that suddenly erupt to the surface, sometimes the origin is psychic absorption of the thoughts and emotions of others, and sometimes the origin is derived from "promptings"of other-worldly forces or entities.

All of these behavioral symptoms reflect the same evolutionary intent: to dissolve all the old barriers preventing a direct contact or relationship with the Universal Source, and to align with a transcendental belief system in order to foster this relationship. In worst case scenarios, the resistance to this process will produce insanity, possession, or total disintegration and fragmentation. In a few, it will manifest as the Jesus syndrome: individuals who are totally deluded and intoxicated by their own egos, and who consider themselves God.

Judgment and discrimination are necessary so that these individuals learn to understand what is happening inside of them. This understanding can only occur through the desire to align with a transcendental belief system or structure in order to realize, and place in proper perspective, the cause behind the manifestation of any or all of these symptoms. In this way, they will learn how to sort out delusions, illusions and fictions versus revelation, truth and divine inspiration. The fear of the unknown will be replaced with the faith and trust to plunge into the abyss of the unknown, and to become resurrected or metamorphosed. In so doing, the fear that some of these individuals have of being dissolved into nothingness or nonexistence will evaporate.

With the bottom line anchored in the universal, some who have positively responded to the prior evolutionary intent will come into this life as natural psychics, channels or mediums. Others will plunge totally into the abyss of infinity with faith and will have an almost exclusive focus upon the Source. A few in this group will have experienced absolute revelations as to the nature of creation. They will experience themselves as Cosmic Beings. Some have realized or felt that "nothing else works" to such an extent that they have or will deny other aspects of themselves that need to be acknowledged and developed — not suppressed. Each individual is unique. Each chart will reveal the prior evolutionary and karmic background that has led to any or all of these conditions.

Others with Twelfth House Plutos will come into this life not understanding why they are here, the Earth being sensed as a foreign environment. These individuals have either spent many other lives elsewhere (other planets), or have spent a great deal of time out of body. A common problem of these types is a frustration at being unable to express to others how they feel and experience themselves.

Some Twelfth House Pluto individuals have had a series of lifetimes of utter withdrawal from the world itself. This withdrawal could have occurred in monasteries, prisons, in the wilderness and so forth. The prior intent was one of self-contemplation within the universal whole. Sometimes this withdrawal was enforced against the person's will through confinement, in order to enforce the same lesson. In cases of forced confinement, the element of karmic retribution must be considered and evaluated as to the reasons why.

All Twelfth House Pluto individuals will have a highly stimulated pineal gland. As Manly Hall points out in The Occult Anatomy of Man, this gland was called the occult gland in other cultures and times. Located above the spinal column, in the middle of the brain, this gland secrets a naturally transcendant chemical substance called melatonin. Just as the ingestion of LSD, for example, will alter the state of consciousness, so too does the pineal gland through the secretion of melatonin. The pineal gland is stimulated by light entering the retina of the eyes. Because this gland is highly stimulated or active in these individuals they can experience, on a cyclic basis, altered states of consciousness. The processs is a naturally spiritualizing one. Yet for those who have not consciously desired to pursue and develop their spirituality, this process can produce the negative psychological symptoms mentioned above.

From a physiological point of view, melatonin produced from the pineal gland sensitizes the entire organism. By sensitizing the brain (consciousness), the individual's neurological (electrical) impulses are sensitized or attuned to "higher vibrations", impulses, or knowledge of a transcendental nature. This substance sensitizes through sharpening all our natural anatomical sensors to any stimulus: touch, taste, hearing, smell, and seeing. For example, a person may become more sensitive to impurities in food. If such individuals do not exercise caution they can suffer problems in their pancreas (enzyme, insulin production), gall bladder, stomach, duododenum, liver, digestive tract, necessary levels of bacteria in the colon and intestines, endoctrine system, and the strength of the astral or etheric body. Emotional, psychic, intellectual or physical stress can produce physiological and psychological problems as can resistance to the evolutionary pull. If the individual is extremely denial-oriented or suppressed, the worst case scenarious promote cancer, tumors, boils, or abscesses.

From a prior-life point of view, all of these individuals have been learning how to balance their need for down time with their need for external activity. This balance is critical. Too much of either can promote emotional distortions of all kinds, loss of perspective, loss of a center of gravity, and psychic distortions. Because it has been so very easy for these individuals to lose touch with themselves, it has also been easy to lose awareness of the need for this balance; of when to retreat or seclude themselves and when not to. Losing touch reflects the dissolution process, the over identification with an aspect of themselves, or the chasing of one fantasy or dream temptation after another. The balance point or rhythm is ever shifting, and never the same. This shifting occurs because these individuals have also been attempting to live in the eternal now, the moment, as it reflects the past and future simultaneously. Thus, it has been and is important to develop an ongoing inner attunement or awareness as to the shifting nature of this rhythm for inner and outer activity, and to act accordingly. If these individuals try to follow this need as it occurs, then they have or will develop a consistent clarity or under-standing at all levels of themselves, and their worldly activities, because they are in harmony with themselves and the universe. The Chinese philosophy of Taoism reflects this necessity in the concept called wu-wei. Simply translated, it means non-action. If the individual develops a conscious awareness as to the need of any moment and acts accordingly, then one's actions are in harmony with what is required in that moment. Thus, there is no action because action implies act-ing upon the moment from an egocentric point of view.

The three most common reactions to this prior evolutionary intent, and the related karmic conditions that occur because of these reactions, can be simply stated as follows:

1. Some Twelfth House Pluto individuals will have rejected or repelled the evolutionary desire through the strength of their egos. Thus, turning their backs to the abyss, they have retreated to the light of the known world; the past. Latching onto something familiar, to tradition, they deny any larger force or source beyond themselves. They become intoxicated with their own egos. Karmically, this reac-tion will produce, on an ongoing or progressive basis, life situations or conditions of increasing powerlessness. Possible manifestations include physical disabilities, situations of confinement, or life cir-cumstances wherein they are karmically blocked from being able to exercise any power at all.

2. Other individuals stand poised on the precipice, unable to move forward or backward more than a few steps either way. They commonly sense that there is a universe and Source to which they are connected, yet still are bound by compulsions and the fear of going too far into it. There usually is a subconscious resistance to submitting to a higher power than themselves. They are simultaneously attracted and repulsed from developing a conscious relationship to this Power. Thus, they practice their own kind of nebulous spirituality. In some the need to develop their spirituality is only acknowledged in moments of extreme crisis when all else fails. As soon as the crisis passes, so too does the need and focus upon their spiritual side. Commonly, these types of individuals will align themselves with work of a human service nature. Karmically, they will experience aliena-

tion, emotional difficulties or disruptions, work related problems, psy-chic disturbances, cyles of utter meaninglessness, emptiness or futility, disillusionment, and divine discontent in order to induce a more active development of the evolutionary intent. Some of these individuals will be highly creative icons who can transform our own vision as to the nature of the world. An amazing number of the world's most gifted actors, writers, and composers have had Pluto in the Twelfth House.

3. These individuals have faced the abyss and, with faith, taken the plunge. In this way they have allowed themselves to be dissolved from all the old barriers and egocentric limitations, and to be totally unconditioned from cultural or societal identifications. They have been reborn as an individual through whom the Source was, and is, expressing Itself. In other words, the center of the individual's con-sciousness has shifted from the wave to the ocean. Becoming at one with the Source, these individuals simply do the tasks and duties that they are asked or destined to do. Some will have very specialized "missions" to perform on the behalf of all of us. Karmically, they are rapidly becoming freed from any further necessity to be on the Earth plane. At some point, they will not return to this plane unless they so desire, or are asked to do so by the direction of the Source and Its agents.

Reactions one and three are not common. Reaction two is by far the most commom. Of course, these three distinct reactions can intertwine as an individual ebbs and flows with the inner and outer conditions of life. Individual evolutionary and karmic conditions will help aid our comprehensive understanding as to how and why each individual has reacted, and is reacting, to this prior evolutionary intent.

In general, then, all Twelfth House Pluto individuals have been attempting to learn how to expand the center of their awareness to encompass the universal whole: to become Cosmic Wholeness, to see themselves and all others as extensions or reflections of the Source of Creation itself. In the Tenth House, the individual learned to be a cultural person with a national identity. In the Eleventh House they learned to break free from this limitation and become inter-national or planetary individuals. In the Twelfth House, they are learn-ing to become Universal or Cosmic individuals.

This process required redefining themselves relative to time and space, culture, groups of like-minded Souls, anything that con-ditioned their sense of separate and personal identity that did not accommodate their timeless and immortal identity in the Source, the Cosmic Sea. The need was to understand the individual part they were meant to play in their various lives, and to allow the Source to express Itself through that part.

In this way, the individual must learn how to let go of everything that pertains to their past in order to prepare, once again, for a brand new evolutionary cycle. This new cycle will be represented by Pluto in the First House. All of us keep going around and around until the Twelfth House archetypal intent is fully realized. It is a process of refinement and progressive elimination of our separating desires, and the karma that these desires generate.

The polarity point is the Sixth House or Virgo. The general evolutionary intent is for these individuals to develop specific and practical mental methods or techniques through which they learn how to analyze themselves. By developing these mental techniques, they will be able to understand how and why they work the way they work, and are the way that they are. In addition, they will progressively learn to see, experience, or witness how one part is linked or connected to another part, and how one part or condition influences the expression of every other part.

Relative to their need to connect themselves to a transcendental belief system, this polarity point now demands that they harness their natural meditational state through specific meditative techniques that will allow them to bring into a sharper and more experiental focus their living connection to the Source. Through these techniques and methods, they can examine with their conscious and rational mind the dynamics within their totality that need to be adjusted, changed, eliminated, or purified because of the blockage that these dynamics are creating.

In addition, the polarity point demands that these individuals commit themselves to some form of work of a practical and useful nature to others. The main theme that must be actualized is one of human service. In other words, they cannot remain in isolated and self-centered pursuits. The form that this work takes does not matter. As long as the theme of human service is met, the form is irrelevant. On the other hand, the type of work should reflect what the individual is here to do; it should not be just "any" work. The individual's right work can be determined by his or her natural evolutionary and karmic condition.

In the East the concept of karma yoga reflects this evolutionary need. Karma yoga means that the individual must identify his or her "right" work relative to his or her natural capacities and tendencies, karmic actions, and attunement to Divine Will. An individual who identifies his or her work in this way simply cooperates and fulfills that work requirement, allowing the Divine or Source to express Itself through the work.

In this way, the work itself becomes a vehicle through which personal knowledge, self-realization, self-purification, devotion and humility are obtained. Work expressed in this way would also benefit all those who were affected by the work itself. What is important is the attitude toward one's work, no matter how grand or insignificant it may be from external standards of judgment. The specific form or kind of work is always in response to the needs of the whole at any point in time.

Work is necessary because, of itself, it is a dynamic that promotes a focus through which these individuals can harness and channel the undefined energies of their Twelfth House Plutos. When one is working, one is involved in an activity that promotes processing or focusing of oneself through that activity. Thus work serves as a mirror or lens in which these individuals can experience, witness, see or analyze all the emotions, moods, feelings, images, states of being, inner dynamics and component parts that surface into their consciousness because of the activity of work. In this way, these individuals can adjust, change, eliminate, or purify components or dynamics that are creating blockage, that are being misapplied or misunderstood, or that are delusive by nature.

In addition these individuals are learning lessons about reality as it is: not what they want to see, not what they blindly want to pretend is happening, not what they want to make reality into based on fantasy, delusions, or naivete. The development of this current evolutionary intent usually demands that these individuals experience cycles of crisis. Crisis brings situations to a head, into sharp relief. The nature and function of crisis is to force these individuals to deal with reality as it is: to see things just as they are, and the actual reasons that it is that way. Crisis can come to these people through the emotional, physical, intellectual, or spiritual bodies. Many will come into this life with subconscious desires or intentions to create crises for themselves. Sometimes this need for crises can be quite compulsive and unconscious. In many cases, these individuals are not aware of this pattern and do not understand why crises seem to keep occurring. Some will assume the attitude of the victim. Others will resign themselves to their "fate." Some will compulsively create crises in the lives of others as they undermine the relationships that connect them to others. They undermine by subconsciously desiring to dissolve the foundation that binds them to others, or they undermine by criticism because they feel others should be punished just as they punish themselves for their own errors. Of course, this type of activity only attracts criticism from those upon whom they project this behavior.

Individuals who have abused power in the past, who have been overly intoxicated with their own egos, or who denied the prior evolutionary intent will experience extreme limitations in relation to finding meaningful work that reflects their capacities. These people will experience reality as it is through the crisis of subservient and mundane work. They will feel as though a big hand is forcefully holding them back. This karmic effect will occur to induce humility, to enforce the awareness of forces greater than themselves. Some of these people will also experience physical problems or disabilities. This form of crisis also enforces analysis as to why this condition exists. Even in such a situation, these individuals are meant to assume some form of service-oriented work.

Some Twelfth House Pluto individuals will produce works of lasting value. These are individuals who have been and are "divinely inspired" in some way, and will serve as examples for others to experience, in order to be helped or transformed in some way by the very nature of the work itself. Many of these types, however, need to learn when to take time off for themselves as many become so dedicated to the work, so self-sacrificing, that they have no time for a life and identity outside their work. By ignoring the intrinsic Twelfth House Pluto need to balance work with rest and withdrawal they can deplete or waste themselves. Some crisis will usually intervene to make these individuals pay attention to this need.

As these evolutionary lessons are put in motion, all Twelfth House Pluto individuals will metamorphose into the essence of humility. This will occur through the inverted pyramid effect wherein the totality of cosmic forces pour through the individual. They will reflect an inner illumination that can light the way for others. This effect can occur in all natural evolutionary conditions, and through any role that the individual is destined to play within the cosmic scheme of things. In its highest manifestations, these individuals can be the living embodiment of the Taoist principle of wu-wei; their actions will be in precise harmony with whatever is required of them at any moment in time.

Common characteristics of Pluto in the Twelfth House or Pisces include: deeply private, not what they seem to be as interpreted in others' eyes, deeply sensitive, take things to heart, amazingly shy at a core level, ultra-emotional and, although you may not know it, can be extremely giving in a silent kind of way. Many deep and unresolved fears, an aura of dreaminess, powerful dreams or never dreams because of exhaustion, naturally psychic.

Famous people with Pluto in the Twelfth House or Pisces:
Teilhard De Chardin
Clara Barton
Christopher Isherwood
Johann Sebastian Bach
George Patton

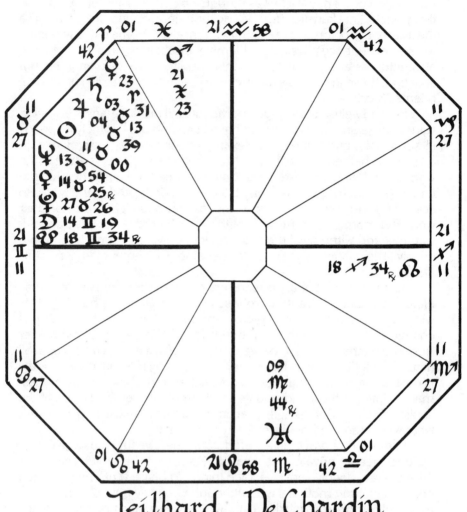

Teilhard De Chardin
Source: Lois Rodden

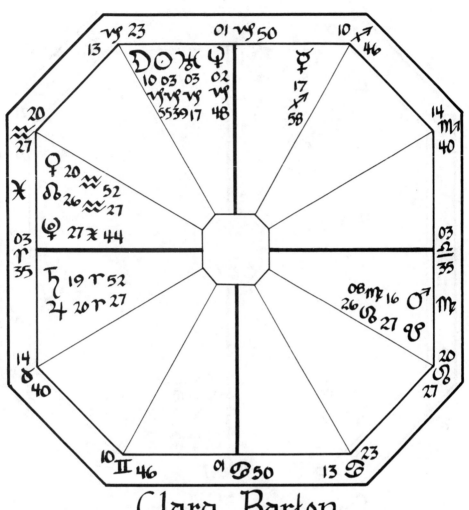

Clara Barton
source: Marc Penfield

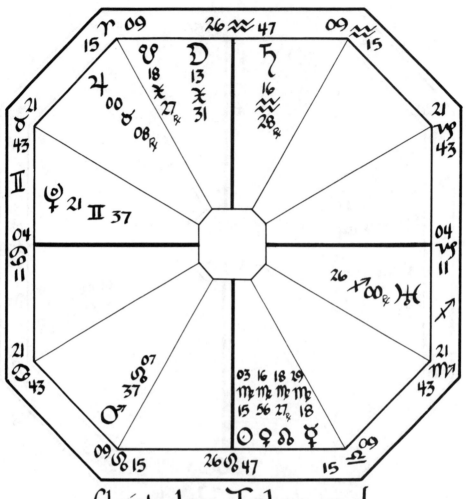

Christopher Isherwood
Source: Lois Rodden

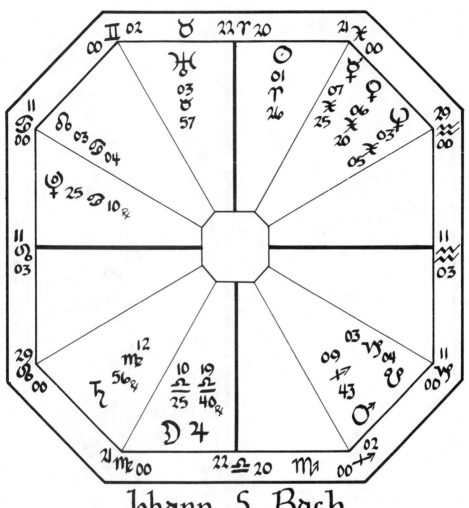

Johann S. Bach
Source: Marc Penfield

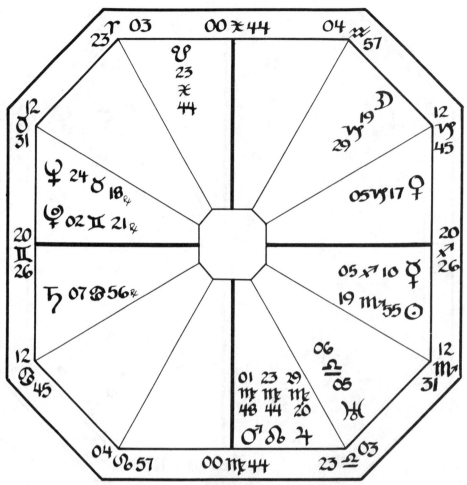

George Patton
source: Lois Rodden

EXAMPLE CHARTS

Two example charts will illustrate the ideas, principles and pro-cedures that we have covered thus far. The specific procedure is as follows:

1. Determine through observation or conversation the evolutionary condition or state of the person in question.
2. Consider the societal, cultural or socio/political context of the individual.
3. Locate the sign and natal house position of Pluto. This will des-cribe the prior evolutionary intent and the area to which the individual will naturally gravitate for unconscious emotional security reasons in this life. Determine the number of aspects that Pluto forms to other planets in order to evaluate the "evolutionary pace" of the person in question. Is Pluto retrograde?
4. Locate the South Node by house and sign to determine the prior mode of operation that allowed the prior evolutionary intent or desires to be actualized. Check for aspects to the South Node from other planets to determine other areas (houses) or functions (planets) used in the past to promote the natal Pluto's prior evolutionary intent. What kind of aspects?
5. Locate the planetary ruler of the South Node by house and sign. This planet describes additional areas in the past that helped facilitate the mode of operation described by the South Node.
6. Determine Pluto's polarity point by house and sign in order to evaluate the current evolutionary intent.
7. Locate the North Node by house and sign to determine the current mode of operation that will be used to promote the development of the current evolutionary intent described by Pluto's polarity point. Check for planetary aspects to the North Node. These planets, by house and sign, correlate to additional functions or areas that will be used to develop the mode of operation described by the North Node. What kind of aspects?
8. Locate the planetary ruler of the North Node by house and sign. This function and area will be the primary facilitator to help actualize the mode of operation described by the North Node.

We will now apply this procedure and the principles described in the first two chapters to two charts. In a succinct and capsulized

239

form they will illustrate the main karmic and evolutionary dynamic in each birth chart, the bottom line, in order to show how the individual's past development conditions how he or she expresses and manifests him or herself in this life.

If you have some difficulty in following these chart analyses, especially in regard to the Nodes, please refer to the descriptions of Pluto through the houses. In Hitler's chart, for example, the South Node is in Capricorn in the Third House. So, you would read Pluto in the Third House and Pluto in the Tenth House. The key is to synthesize these explanations, and to apply them to the intrinsic meaning or principle described by the South Node, a mode of operation that fostered the development of the prior evolutionary intent described by Pluto's natal house placement. Similarly, Pluto is in Gemini in the Eighth House in Hitler's chart. So, read Pluto in the Third House and Pluto in the Eighth House and synthesize the two in order to understand how the prior evolutionary intent of the Soul was developed and actualized.

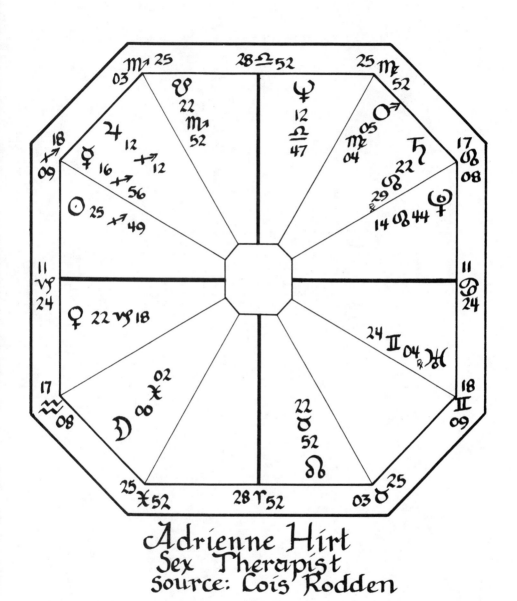

Adrienne Hirt
Sex Therapist
Source: Lois Rodden

ADRIENNE HIRT: SEX THERAPIST

Profiles of Women by Lois Rodden describes Adrienne Hirt as follows. "After a series of youthful jobs, a seven-year marriage that terminated in 1967, and a daughter, she became a belly dancer. She liked the sensual spontaneity of the dance, club work, personal appearances, and class teaching. In 1973 she attended a UCLA class on sensitivity and 'something clicked' inside of her. It was another year before she was able to take Sylvia Kak's sensitivity workshop to learn the techniques of holistic body training."

"She felt a compelling need for in-depth relationships with others and satisfied that need through her career. In 1975 she became a surrogate sex partner, working through the recommendation of psychologists with about seven clients a week. She sees her clients once a week for as little as four months, or perhaps up to two or more years. 'This is healing and teaching work,' says Hirt. 'A man with a problem in sexual dysfunction needs to learn, not intercourse, but intimacy. I'm helping, and I feel good about it' ".

Based on the above information, we can safely assume that Adrienne is in the individuated evolutionary condition or state. She is white, and her socio/political context is the United States and democracy. She has six aspects involving Pluto including the squares to the North and South Nodes. Hirt is attempting to achieve accelerated growth and to do a lot of work on herself in this life. With Pluto retrograde, she will do it in her own way. Her entire life will be characterized by cycles of deep metamorphosis.

With Pluto in the Seventh House retrograde in Leo, the South Node in Scorpio in the Tenth House (prior mode of operation), and the facilitator being Pluto itself (the planet ruling the South Node), and Pluto square the South Node, we can make some simple yet fundamental observations about the evolutionary and karmic conditions pertaining to her past that set up this life's evolutionary needs.

These symbols demonstrate that there have been conflicting desires and needs carried over from the past. On the one hand, Pluto describes her prior need for intimate relationships with others in order to define her own individuality through counterpoint-type comparisons or evaluations. It describes her prior need to learn lessons pertaining to giving, equality and relativity. It describes her Leo need and desire to be recognized and considered special by others. It describes her dependency on others for these lessons, and her compelling need to relate to others deeply and intimately. Thus, coming into this life Adrienne married very young (she was born in

1947) and had a child. This experience reflects the natural gravitation of the Seventh House Pluto in Leo (prior patterns in identity
association). On the other hand, she has also needed to actualize her
own reason for being through a career: South Node in the Tenth
House. After her marriage ended, she became a belly dancer, per-
forming for audiences and teaching (Pluto trine Mercury and Jupiter
in Sagittarius in the Eleventh House). No doubt she could do this
very well. The need for attention and recognition was now expressed
through public relationships linked to her career (Pluto in Leo in the
Seventh House, South Node in Scorpio in the Tenth House).

One karmic and evolutionary problem here is the famous
Scorpio/Pluto trap of either/or. In this case she had to choose between a consuming relationship and a consuming career. With the
South Node in Scorpio, square Pluto retrograde in the Seventh
House, and Saturn retrograde in the Eighth, she is reliving conditions
pertaining to her past. The first marriage was the bottom line to
which she naturally gravitated in order to find her own sense of
individuality and personal completion. In so doing she fulfilled deep-
seated security needs. Yet, she also created this situation to reawaken
the evolutionary conflict relative to her career needs.

Thus, in becoming consumed by the relationship and responsibilities of being a parent (South Node in the Tenth), she created a
situation of confinement and limitation. This situation ignited the
inner volcano of her own career needs, and the evolutionary desire
and intent to develop her own individuality on her own terms (Pluto's
polarity point with respect to the First House). She had to break free
from the weight and obligations of meeting her partner's needs and
expectations in order to actualize her own special and unique destiny. Rather than depending on a mate, friends, or parents for answers
to the penetrating questions that she would naturally ask, she had
been learning to answer and supply her own needs, to achieve complete self-reliance, and to establish emotional security from within
(North Node in Taurus in the Fourth House).

Because Pluto is also square the North Node, and Venus the
ruler of the North Node is sextile to the South Node and trine the
North Node, this life is not the first one in which she has
attempted to learn these lessons of self-reliance and inner security.
These symbols suggest that so much karmic residue has been built
up over so many lifetimes relative to intimate and sexual situations
that the need to relive and fulfill that karma is overwhelming. The
temptation to "skip steps" is indicated by the square from Pluto to
the Nodal Axis; in other words, to throw off the necessity to resolve
the issue of learning how to be in relationships while at the same

time developing and actualizing her own identity and career. The "skipped steps" translate into an either/or approach to this issue. In this life they must be recovered and fulfilled through a "both/and" approach to her life necessities. This past karma is intense because these symbols suggest that she has rejected commonplace solutions to her questions. Clearly she has desired to look into the depths of life and ask why life, why death. She likely became aware of forces larger than herself many lifetimes ago.

On this basis she has initiated many relationships with other individuals whom she felt possessed something that she needed in order to answer her questions and fulfill her needs. These symbols suggest that these relationships were manipulative and unequal by nature, bringing clashes or confrontations revolving around questions of authority and control. In some cases she would attempt to be the defining and controlling person, in other cases she was defined and controlled by another, and in some instances she and the other individual would attempt to alternate these roles within the relationship. These symbols suggest that these relationships were only maintained for the duration of her need or the other person's need. They suggest hypnotic and sexual attractions that were compulsively followed to find out what the attractions were about. They suggest many terminations (leavings and partings) and unresolved issues (karma) pertaining to many people with whom she had been involved over many lifetimes. They suggest a subconscious guilt pattern (South Node in the Tenth House square Saturn and Pluto) as a result. In addition, these symbols suggest many emotional wounds and scars for her and for many of those with whom she had been previously involved: wanting to trust, yet being afraid to trust at the same time. Consequently, she was karmically destined to relive many of these situations in this life; not to skip steps.

From a prior-life point of view, she has been attempting to break free of the chains of such relationships in order to establish her own career and identity (Venus in Capricorn in the First House, sextile the South Node, trine the North Node, inconjunct Saturn retrograde, and inconjunct Uranus in the Sixth House). These symbols suggest her need to actualize her prior lives in new ways, to be in relationships in new ways, and to rebel against societal and parental customs. Yet, Pluto is in Leo in the Seventh House. This placement correlates to her need to be emotionally recognized and accepted by her parents, society and lovers. With Pluto square the Nodal Axis, this recognition could not come because of her ongoing desire to achieve self-reliance, inner security, and to establish her own authority, values, beliefs and identity, and the lifestyle that would reflect these factors.

Thus, this conflict has created situations wherein her parents, lovers and society could not give her what she needed. In other words, she had to draw parents who forced her in upon herself through their inability to give her what she needed emotionally. Over many lifetimes, this situation created displaced emotions wherein she would attempt to have those emotional needs met through lovers. This dynamic set in motion deep subconscious expectations that were projected onto her partners (ruler of South Node in the Seventh House).

The resulting confrontations produced one emotional blow after another that would finally lead to separations. Because outer reality is a metaphor for our inner reality, she would also attract individuals who also have displaced emotional problems. The effect of confrontations leading to separations would lead into enforced self-sufficiency at all levels, both for her and her partners (Pluto square the North Node, Venus in Capricorn in the First House trine the North Node, Moon in Pisces in the Second House trine the South Node and septile the North Node). In addition, these experiences forced her to identify and rely upon her own resources to effect physical and emotional security, and to learn to accept the responsibility for her own actions (why is this, or why has that happened).

Implicit is an element of karmic retribution. With the Moon in Pisces in the Second House septile the North Node, trine Uranus in the Sixth House, and Uranus inconjunct the South Node and sextile the Eighth House Saturn, which is square the South Node, a career or work that is oriented to human service, psychology, and sexuality is a natural outlet. Such work allows her to "feel good about herself," which implies that she is beginning to come to terms with the guilt from the past. Sensual, sensation-oriented work allows for physical and emotional healing within herself, as well as for those with whom she works. Her initial interest in sensitivity training allowed her to deal with the old wounds and scars from the past so that she could work with others in the same way.

Relative to issues carried over from the past that must be relived, and also to fulfill the element of retribution, she has succeeded in creating a new career that meets her compelling need for in-depth, intimate relationships and through which she heals the emotional hurts of others. Also, she is in a position of control, as opposed to being controlled by another (South Node in the Tenth House in Scorpio, Pluto in the Seventh in Leo, North Node in Taurus in the Fourth House, the planetary ruler, Venus, in Capricorn in the First House). She is the authority figure in this work. Also the work is radical and new enough that the potential for criticism (Uranus in the sixth inconjunct Venus in Capricorn, inconjunct South Node in Scorpio, sextile Saturn in the Eighth, trine Moon in Pisces) will ensure that her motives are sufficiently pure.

Pluto's trines to Jupiter and Mercury in Sagittarius in the Eleventh House and Scorpio (Pluto rules Scorpio) on the Eleventh House cusp correlate to the multitude of sexual contacts that her work involves, allowing her to become "impersonally/personally" involved with these men (Eleventh House). This dynamic enforces necessary detachment because her clients leave after their period of treatment is over. She must let them go in order to maintain her own independence and self-reliance at all levels. Karmically, her work provides renewed contact with all those with whom she has unresolved relationships from other times. Here is an opportunity to resolve difficult and painful issues left over from previous cycles.

The South Node and North Node in the Tenth and Fourth Houses clearly demonstrate that an old evolutionary and karmic cycle is coming to a close while another is in the process of unfolding. Again, Pluto square the Nodal Axis demands a reliving of these issues in order for this old cycle to culminate so that she may finally be freed from it once and for all. With Pluto applying to the North Node, these experiences must occur so that complete self-reliance is realized on all levels.

Having the buffer of psychologists is also good because it provides a selection or discrimination process (Uranus inconjunct the South Node, inconjunct Venus, sextile Saturn). The Pisces Moon opposing the Virgo Mars in the Eighth House suggests that the discrimination process has not been very well developed before. Psychic impressions, yes, but the anaysis and doubt of the Virgo Mars can, and has, confused or overridden these impressions. This aspect pattern suggests confused or unrealistic evaluations, inability to recognize reality or motives, and falling for the sad stories of others. Thus, it is good that she trusts (Moon) the evaluations of external, objective authorities (south Node in the Tenth, Uranus in the Sixth). On the other hand, the Moon/Mars pattern will allow her to penetrate to the core of another in order to identify where they are pinned and what the basis of their emotional/sexual difficulty is. Mars in Virgo in the Eighth in opposition to the Moon in Pisces in the Second further indicates a personal resource that she can use to sustain herself on a monetary plane is linked to her work as a sexual surrogate.

With the Sun in Sagittarius in the Twelfth House inconjunct the North Node, teaching and writing can develop as natural outlets to expand her career and as a vehicle for self-integration and creative application of her purpose. Linking this dynamic to a transcendental or metaphysical belief structure will allow her to develop the total awareness through which she can come to understand this entire process. In addition, she must learn how to respond to her need for down time and self-contemplation so that she does not deplete herself and lose contact with her own center of gravity. In this way, the cyclic metamorphosis of her personal sense of individuality, which will expand into larger and larger concentric circles, will develop with continual clarity of self-understanding.

This lady had struck out on her own (Pluto polarity into the First House), and fulfilled her own creative purpose (Pluto in Leo) through linking it to a socially relevant need (Leo polarity in Aquarius). At some point it will become necessary for her to fulfill relationship and karmic needs through creating a sustained and committed partnership that allows for her career activities to occur in a supportive and non-threatening way. Clearly, this need demands a partner who is self-reliant and secure (the ruler of the Seventh House Cancer cusp being the Moon in Pisces in the Second House). If she decided just to teach, lecture, and write, or change careers altogether, she would need a partner with an enlightened attitude toward her past activities. In these ways, Adrienne is fullfilling her karmic and evolutionary desires and needs.

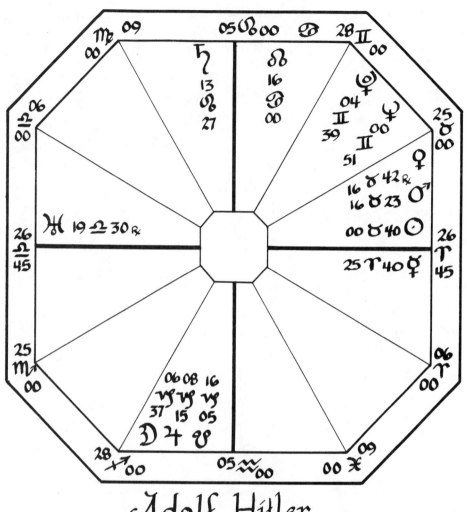

Adolf Hitler
Source: Marc Penfield

ADOLF HITLER

Based on the history of Hitler's life, I feel that he was in a transitional evolutionary state; a transition from the herd to individuated state. This evaluation is based on the observation that he thoroughly understood the ways of society, yet, in his own way, tried to break from the past by developing new approaches and ideas in order to fundamentally restructure his society. Remember that those who have been in the herd state for a long time and have learned how the "system" works, and who are moving toward the individuated state, can be individuals who develop the capacity to lead or control the "herd." Hitler's socio/political context was, of course, a Germany crippled by runaway inflation and a national sense of powerlessness. Hitler has four aspects to Pluto, two of these aspects are inconjuncts, one a sesquiquadrate, and one a conjunction to Neptune in the Eighth House.

With Hitler's Pluto in the Eighth House conjunct Neptune in Gemini, he had a series of prior lives in which he was attempting to learn the lessons pertaining to power and powerlessness. This suggests his evolutionary need to learn the limits of power. These lessons imply an awareness coming into this life of forces larger than himself. Pluto's inconjunct to the Capricorn Moon and Jupiter in the Third House, Jupiter and the Moon being conjunct the South Node, suggests his prior-life need to wield social or national power. These aspects correlate to his innate ability to understand, intellectually, the basis of social structures. Jupiter and the Moon also inconjunct to the Tenth House Saturn in Leo, the ruler of the South Node, implies that he had in fact achieved positions of power and leadership in the past.

The means to achieving this power were manipulative and divisive by nature (Pluto conjunct Neptune in the Eighth House). Given all the inconjuncts in this chart, and Pluto's conjunction to Neptune, the prior-life awareness of power and powerlessness revolved around persecution and the persecutor, conspiracy and the conspirator. It would seem to me that Pluto conjunct Neptune in Gemini in the Eighth House also suggests that he had studied the writings, philosophies, and ideas of others pertaining to utopian or idealistic social systems (Jupiter on the South Node inconjunct Pluto and Neptune.

The Capricorn signature in the Third House inconjunct the Tenth House Saturn in Leo confirms this inclination, and also suggests that the nature of these idealistic or utopian social systems would be structured in a "pyramid type way" (Saturn in Leo). Beyond

251

idealism, this dynamic would, of course, translate into a dictatorship, or an authoritarian regime that imposed an "ideal" upon the masses. From a prior-life point of view, these same symbols suggest delusions of grandeur. They also suggest that Hitler had traveled far and wide in other lifetimes, seeking out teachings and ideas from many different cultures and sources. It seems likely that his compulsion to force his ideas down other people's throats got him into a lot of trouble in other times.

With Uranus in the Twelfth House retrograde in Libra, square the South Node, opposed Mercury (itself square the South Node), and Uranus sesquiquadrate Pluto, it also seems that he must have experienced lifetimes of persecution because his ideas were considered crazy or too radical. It would not be unlikely that he had been incarcerated in an insane asylum. With Pluto in the Eighth House connected to Uranus, and Uranus inconjunct Mars and Venus, opposed Mercury in Aries in the Sixth, it would seem likely that the accumulating rage built up over many lifetimes would translate into a desire for revenge, vindictiveness, and grand schemes to pay back the "persecutors."

So Hitler would come into his last incarnation with this type of prior-life pattern. The ruler of the North Node, the Moon, conjuncts the South Node, and Uranus and Mercury both square the Nodal Axis. These aspects indicate the need to relive the karmic and evolutionary dynamics of the past. He would naturally gravitate to these prior orientations. He chose a father who was extremely judgmental and authoritarian (Saturn in the Tenth in Leo square Venus and Mars in Taurus in the Seventh, inconjunct the Moon, Jupiter and the South Node in Capricorn). This pattern reflects his own need to judge himself for prior sins, as well as to experience an element of karmic retribution through judgment by his father because of his own judgmental patterns from the past. He chose to be born into a culture that was experiencing runaway inflation and powerlessness on an international level. Hitler's own early life was colored by a sense of insignificance and powerlessness.

Hitler's intrinsic desire for recognition and power finally put him in jail for political rebble-rousing activities. He had been reading and studying the philosophies and ideas of Neitzche, Hegel, Marx, and Eastern mysticism among others. As these ideas fermented through time he finally synthesizd them into his own system, the philosophy that would lead to the formation of the National Socialist movement. The books and ideas that he was studying represented the transformative symbols that allowed him to effect a metamorphosis of his sense of powerlessness, from a societal point of view, into a state of

power with respect to the synthesis of these ideas into his own system through the doctrine of Nazism. This synthesis of ideas was
directly linked to his perception of Germany's problems. He created
a Master Plan, a new philosophy or vision to solve those problems.
This vision or plan was a projection of his own evolutionary need to
learn self-reliance and self-sufficiency, and to identify and use his
own resources to sustain himself (Pluto polarity into the Second
House with respect to the North Node in Cancer in the Ninth
House.)

This plan also projected his own evolutionary need to develop
internal security (North Node in Cancer), and to secure his own borders so that he could be in control of his own life. Germany at that
time was at the mercy of other nations, had no real sense of national
purpose, and was not in control of its own destiny. The point is that
Hitler's own development was inextricably linked to the karmic necessity
to relive a life of social position, leadership and total authority. With
Saturn in Leo, he already knew how to take destiny and shape it by
the strength of his own will. He linked this innate ability to a national
vision that he communicated to the German people: to take control
of destiny on a national level (North Node in Cancer reexpressed
through the South Node in Capricorn in the Third House conjunct
the Moon and Jupiter, inconjunct the Tenth House Saturn in Leo).

He borrowed ideas from Neitzche and developed them into the
concept of a master race based on genetic purity (Uranus in the
Twelfth square the Nodal Axis and opposed Mercury in the Sixth with
respect to the bottom line of Pluto in the eighth conjunct Neptune in
Gemini). The Eighth House, Pluto, and Scorpio all correlate to
human genetics. The Twelfth and Sixth Houses correlate to purity. In
Also Sprach Zarathustra, Neitzche expounded a philosophy of super-
men who descended from heaven to lead the masses. He also wrote
a book called *The Will to Power,* suggesting one could shape destiny by the force of one's will. Hitler believed this idea could be
applied to the national will. He communicated and projected these
concepts from an idealistic point of view, and adopted symbols for
the masses to identify with and rally around. The symbol of the swastika originated in Tibet as a representation of the four cardinal points
emanating from the source. Hitler used the swastika to symbolize the
axis of power that would emanate from Germany. To the German
people it served as a rallying point or symbol to unite the national will
to take charge of their individual and collective lives once again.

Hitler had only to find a scapegoat; the Jewish race. He had to
use this tactic to reinforce his idea about regaining control of Germany's destiny. Thus, the Jews were blamed for every problem. They

were considered as an infectious disease, contaminating Germany from within. One of his main arguments to prove this assertion to the German people was based on economic issues because the Jews owned many businesses and to a large extent controlled the banking system (Pluto in the Eighth House conjunct Neptune in Gemini). By isolating and persecuting an entire subgroup of people within Germany (Uranus, groups, in the Twelfth House in Libra), he attempted to regain control and power for himself and the nation (Pluto in the Eighth conjunct Neptune with respect to the South Node in Capricorn, Pluto sesquiquadrate Uranus).

These ideas, visions and philosophy restate the insanity and radical nature of his ideas from other times. Yet the masses accepted these ideas because of the situation (cultural context) in Germany at that time. Hitler was able to communicate his ideas forcefully and effectively because of prior-life development of his oratory and communication skills (Nodal Axis in the Third and Ninth Houses, Pluto conjunct Neptune in Gemini in the Eighth House). His power of communication was so well developed that he mesmerized everyone. He created a national trance because his powerful persona and his ideas penetrated the Soul of the masses. With Uranus connected as it is to these dynamics, Hitler also became a national icon who symbolized to the masses what they themselves wanted to become.

Hitler devised many symbols to evoke national identity. In effect, he was manipulating the consciousness and minds of the masses. He attempted to make Germany the center of power upon which other nations and people would be dependent. In so doing, Germany could then manipulate the national resources of all nations, and could channel those resources through Germany to make it even more powerful (Pluto polarity into the Second House, reexpressed back into the Eighth House). In this way Hitler himself would be ever more powerful.

Hitler's policies drew horror-stricken reactions from other nations as the reality of these policies became revealed; his racist theorizing led to the slaughter of millions of Jews, Gypsies, Communists and other "undesirables." The illusory national dream turned into the reality of a national nightmare (Uranus opposed Mercury from the Twelfth to the Sixth Houses with respect to the Nodal Axis). As the end approached, Hitler manifested his inherent emotional and psychic instability. Assassination attempts by his own officers, i.e. his "family" (Moon and Jupiter in Capricorn inconjunct the Eighth House Pluto and Neptune), and by other nations (Uranus square the Nodal Axis, sesquiquadrate Pluto) reflected his past life experiences of persecution and downfall. Becoming progressively irrational, he initiated strategies and policies that only hastened his own and Germany's downfall. He harbored a subconscious death wish in order to atone for the guilt that he felt deeply within. The final act of suicide (Pluto in the Eight conjunct Neptune) was effected in order to avoid the humiliation that would come at the hands of others, as he would have been held accountable for the crimes against humanity committed in his name (Pluto in Eighth sesquiquadrate to the Twelfth House Uranus).

It is interesting to note that Hitler came to power in the same year as President Franklin D. Roosevelt. They both died in the same year, the same month, and within a few days of one another. The forces of good and evil seem to be reflected in these two national leaders. We can see Hitler's problems with Roosevelt and democracy reflected in his Ninth House North Node in Cancer. This symbol stands for that which is morally right from an intrinsic or ultimate point of view. Hitler was destined to play the role that he did owing to the karma he built up over successive lifetimes as reflected through his desires. Interestingly enough, Roosevelt's Moon in Cancer is conjunct Hitler's North Node. His Saturn is Square Hitler's Saturn, and Hitler's Pluto

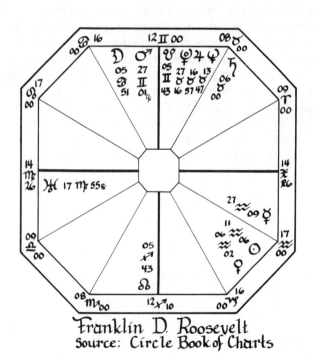

Franklin D. Roosevelt
Source: Circle Book of Charts

and Neptune conjunct Roosevelt's South Node. Even though Roosevelt supposedly manipulated events (Pearl Harbor) in order to justify entering a war, the United States had to enter the war for much deeper reasons than anyone realized at the time — or pehaps don't even realize now. Hitler obsessively believed that Roosevelt was the primary obstacle to achievement of his ambitions. Hitler thought Roosevelt's death was the "sign" that would allow him to prevail. A few days later, as the "larger forces" surrounded him, he took his own life.

There will be a full chapter on transits, but I would like to point out some remarkable symbolism in regards to Hitler and Roosevelt. Roosevelt passed away on April 12th, 1945. On that day Pluto was trine his North Node, sextile the South Node. Pluto was also opposed his natal Fifth House Aquarius Sun and Mercury, and square his Eighth House Saturn in Taurus. In addition, Saturn was transiting through exact conjunction to his tenth House Moon, and inconjunct the Sun and North Node. As if these aspects were not enough, transiting Sun, Moon Mercury and Venus (all in Aries) were in his Eighth House. Jupiter was conjoining his First House Uranus in Virgo.

Hitler committed suicide on April 29th, 1945 as the transiting Nodal Axis was approaching a return to his natal nodes in Cancer and Capricorn (the ending and beginning of old and new cycles). The transiting South Node was inconjunct his natal Tenth House Saturn. Pluto was approaching a conjunction to his natal Tenth House Saturn, and inconjunct natal Moon and Jupiter. Neptune was in an exact transiting trine to his natal Eighth House Pluto. Transiting Saturn was approaching conjunction to his natal North Node, opposed the South Node, and in exact opposition to his natal Moon. Obviously, he saw the handwriting on the wall and committed suicide to escape the avenging forces that were closing in on him.

CHAPTER THREE

PLUTO IN ASPECT TO OTHER PLANETS

A Word on Pluto's Specific Aspects

In this chapter we will be discussing Pluto's aspects to specific planets. Please keep in mind that the number of aspects that Pluto forms will determine the individual's evolutionary pace in this life. In fact, any aspect between Pluto and another planet suggests that the experiences associated with that planet and the house that it falls in have previously come under Pluto's evolutionary impulse in order to transform or reformulate the expression of that function or area. The type of aspect correlates to the approach or attitude by which the individual dealt with the evolutionary impulse to eliminate blockages or barriers preventing further growth. Each person will have a unique approach to this process. The prior-life approach will be reflected in the individual's attitude and approach in this life. Each birthchart is a composite of many interrelated personality dynamics that constitute the whole nature of each person. How one person approaches a Pluto square Venus, for example, will depend on that person's total nature and general approach to life. Even though the archetypal meaning of Pluto square Venus is the same in all cases, the response to this archetypal impulse is different in all cases. This point is further magnified when we consider the four natural evolutionary conditions or states into which individual has evolved.

In the last analysis, the individual response to this evolutionary impulse will be reflected in one's values and beliefs. Where one person may value the necessity of continual growth, welcome it even, another may desire an easy or nonstressful approach to this life. The latter individual would tend to resist necessary change, while the former would welcome and understand change as necessary. Sociological factors must also be considered. These points are important to understand because the traditional meanings associated with aspects, i.e. a trine equals ease, a square equals stress, do not manifest themselves in fixed or unalterable ways in all birthcharts. The archetype of the aspects is

257

constant, i.e. a square equals stress, yet how each individual responds to and manifests "the stress" is different in each case. As an example, some individuals would not interpret the Pluto square Venus as stress at all. They may respond to this evolutionary impulse as a necessary refor-mulation or metamorphosis of their ways of relating to themselves and others. Their attitude toward this cyclic metamorphosis would be one of understanding and cooperation (nonresistance). Such an attitude would not mean that the experiences of those metamorphic cycles would be less intense or difficult, but that the individual's approach to and under-standing of the process would be less emotionally and psychologically stressful than for other individuals who resist the evolutionary process.

Similarly, the traditional interpretation that trines, sextiles, and non-stressful aspects mean ease of understanding does not always apply. In some people the nonstressful aspects to Pluto poduce an ease of *resis-tance* to the evolutionary impulse. In other words, it can be easy to main-tain prior or existing compulsive patterns associated with the affected planets and houses in the birthchart. In these cases, the nonstressful aspect turns into stress because of the tendency to resist the required changes. It is easy to fall back upon old patterns of compulsive behavior and to use these old patterns as explanations or rationales to justify (block) not only what is happening on an experiential basis, but also to thwart necessary change. In other cases, the nonstressful aspects do produce an ease of understanding of evolutionary necessity. Each case is unique. We must take into account the whole picture of each individual's evolutionary and karmic condition, as well as their sociological context.

In considering Pluto's aspect to other planets, as well as Pluto's manifestation in houses and signs, it is important to remember that we are dealing with an essentially unconscious process. The desires, motivations, intentions, and needs that originate from the Soul (Pluto) dictate our conscious thoughts, emotions, moods, feelings desires, res-ponses, attitudes and approaches to life. Again, how many of us are aware of that which originates from our unconscious depths? I have found over many years of astrological counseling that unless an individual is relatively developed along spiritual or metaphysical lines, the aware-ness or ability to comprehend or understand these unconscious dic-tates is simply nonexistent. Many people react mechanically or compulsively in a trance-like way to these unconscious commands. They repeatedly and habitually approach certain aspects of life in the same way over and over again. No matter how problematical or difficult this compulsive behavior may be, it does constitute their unconscious sense of emotional security — it is what they know how to do from prior-life actions and development. The degree of resistance to the necessary evolutionary changes will determine the kinds of evolutionary experiences that an individual will have: cataclysmic or slow but steady growth. The evolutionary pressure and process symbolized by Pluto (the Soul) teaches us that we do have free will and choice, and that our life conditions, experiences and orientations are reflections of the desires that we have. As counselors working with individuals who are compulsively and habitually approaching certain areas in their lives in the same way over and over again, we can help faciliatate this awareness for them by pointing out the basis of their compulsions, security patterns, and the desires associated with them.

Pluto's house position and aspects to other planets demonstrate exactly what areas within each person have previously come under the evolutionary process. The kind of aspect demonstrates how the evolutionary processss was working. In general, the stressful aspects, i.e. conjunction, square, sesquiquadrate, inconjunct, semi-square, and opposition produce a tremendous and intensified concentration of evolutionary energy (desire) to metamorphose or reformulate the expression of whatever planet (behavior) that is being aspected by Pluto. This process usually has not been completely fulfilled or resolved in prior lives. Thus, these functions (planets) will continue in this condition in this life. In general, as a rule of thumb, if the orb of the aspect indicates that the planet has not reached an absolute degree, i.e. an exact aspect (Pluto at sixteen degrees Leo square Venus at sixteen degrees Scorpio), then two conditions could exist:

1. If the orb of the planet is before the exact aspect, it tends to indicate that this function has recently come under an intensified evolutionary impulse, and is still attempting to fulfill these evolutionary requirements. It is unresolved because it is a relatively new condition.

2. If the orb of the planet is after the exact aspect, it tends to indicate that this function has been dealing with this evolutionary impulse for quite a while, and is in the process of completing or fulfilling those prior evolutionary requirements.

The distance in degrees from the exact aspect will suggest how "new" or "old" this process is. In general, stressful aspects have created (and will create) a situation wherein the individual cyclically experiences relatively intense stagnations, limitations, resistance, or emotional shocks relative to the planets Pluto aspects and their house placements. Cyclically these areas are forcefully reformulated, redefined, or reexpressed in some way — with or without the individual's cooperation. In other words, the stressful aspects that Pluto forms to other planets demonstrate preexisting patterns of compulsive behavior that block further growth with respect to those functions and experiences. These aspects can also indicate where, and for what reasons, the individual has difficult or negative karma to play out in this life as a continuation of prior-life actions. Rarely does the individual understand the reasons why, or the basis of these of cyclic metamorphoses as they are occurring. The metamorphic experience can be very difficult, emotionally intense, and experienced in such a way that the individual feels as if there are "larger forces" controlling or manipulating the experience itself. The person may feel powerless to prevent or control the experience. Only after the experience is over, or in the final stages, will the individual, in hindsight, understand the meaning of the enforced change — the nature of the experience itself.

The conflict in stressful aspects revolves around the confrontation between the security associated with the old pattern of behavior and orientation, and the evolutionary intent to metamorphose those patterns into new ones. The degree of difficulty or stress is in direct proportion to the resistance that the individual manifests toward the necessary change, resistance based on the fear of the unknown: insecurity. These confrontations occur to make the individual aware of his or her motives, intentions, needs and desires that have created and determined the pre-existing patterns and orientations. Self-knowledge and growth are the result.

In general, the nonstressful aspects, i.e. semi-sextile, septile, quintile, and trine have produced and will produce a relative ease in understanding the dynamics or issues pertaining to the past and future as they relate to necessary growth. The planets in nonstressful aspect to Pluto allow for a relatively easy transition between the past and the future at each moment in the individual's life. Such aspects promote steady and consistent noncataclysmic evolution. The stressful aspects promote a build-up of emotional energy (like water in back of a dam, lava beneath a volcano) that finally explodes (or implodes) to utterly change the behaviorial function that was producing the block. Nonstressful aspects allow for "knowing why" as the continuing evolutionary process is occurring. Even if the individual resists the evolutionary process via the nonstressful aspects, he or she will know why it is happening at some level. If this resistance finally leads to experiences that are necessarily difficult in order for the person to grow, he or she knows why it is happening in that way, as it is happening. The person may not like it, but will understand it.

These nonstressful aspects can also make the individual aware of what must be done to effect change, to grow and evolve. Yet, because there is a lack of stress the individual may not act upon these ideas, and may not implement or pursue the necessary directions or experiences, but maintains the old and comfortable ways of being. Or the individual can make half-hearted attempts to change, or go only a certain distance toward implementing the required changes. At some point, if this lazy approach continues, nonstressful aspects can become stressful either in this life or a life to come. Nonstressful aspects can turn stressful in this life through progressions, transits, or solar returns. Stressful aspects can become nonstressful if the individual cooperates with the evolutionary requirements and necessities. Again, the orb consideration will tell us if this process is relatively new, or if it is relatively old and in the final stages of completion relative to the existing evolutionary cycle.

The specific nature of an aspect within these two broad categories will determine how the Plutonic metamorphosis takes place. At this point it may be useful to explain how each specific aspect works through capsulized summaries of how each aspect works within the total cycle: from zero to three hundred and sixty degrees. For complete information on aspects I would suggest you read *How to Interpret Your Aspects* by Robert Jansky (Astro-Analytics). Also, I have a forthcoming book titled *Phases, Aspects, and Key Planetary Pairs*, a book I was writing with Jansky before he died.

As you read through these descriptions keep in mind their linkage to the current evolutionary intent or condition of Pluto, i.e. Pluto in waxing trine to Mercury. The meaning of each specific aspect should be linked to the evolutionary intentions in order to be properly applied. Again, the evolutionary intentions are described by Pluto's polarity point by house and sign, and the mode of operation described by the North Node with its planetary ruler facilitating the process through its own house and sign locality.

Waxing aspects range from zero to one hundred and eighty degrees, and waning aspects from one hundred and eighty to three hundred and sixty degrees. To understand whether an aspect is waning or waxing simply determine the number of degrees that separate Pluto and every other planet that it aspects. Using Pluto as your stationary or reference point, count in a *counterclockwise* direction to determine the degree of separation, beginning with *Pluto* equalling zero. The only exception to this procedure involves the *Sun*. Because the Sun is the center of the Solar System, we must use it as our stationary or reference point when calculating aspects between the Sun and another planet.

Waxing Aspects

1. **0° — Conjunction:** The planets unite in function so that a new evolutionary purpose and cycle can begin in an instinctual way. The new cycle or purpose will be projected spontaneously, instinctively, without any egocentric awareness. It is pure, unchecked expression or action. Random experiences will be initiated in order to begin the process of self-discovery of this new evolutionary purpose.

2. **30° — Semi-sextile:** The emergence or formation of a conscious

and egocentric identification with the new evolutionary purpose or cycle. Random action narrows as the individual begins to sense specific directions or experiences that will allow realization or discovery of what the new evolutionary purpose or cycle is about.

3. **40° — Novile:** The conscious and egocentric awareness and identification of the new evolutionary purpose begins a gestation process wherein subjective growth towards the development or actualization of the new purpose begins by giving it personal or individual meaning. In other words, the individual becomes aware of moving in new directions and that these new directions are highly personal in nature. The sense of self-discovery is heightened and the narrowing of random experiences continues.

4. **45° — Semi-square:** The new evolutionary purpose intensifies as the individual struggles to establish and make real the purpose in an individual way. This intensity is due to the fact that the individual is attempting to pull away from all past conditions that may bind him or her to an old order, pattern, or regime.

5. **51° 25' — Septile:** The individual associates the new evolutionary purpose or cycle with some form of personal or special destiny. Experiences and actions are initiated in order to realize or discover what this special purpose or destiny is. The action taken can be sporadic and confused, or clear and consistent depending on the planet aspecting Pluto. Uranus could produce sporadic action for example, whereas the Sun could produce clear and consistent action.

6. **60° — Sextile:** The process of consciously understanding the new evolutionary cycle and purpose occurs through contrast and comparison. The individual must isolate him or herself from the impact of the external environment in order to realize and discover from within that which is uniquely new and individualistic about him or herself. Action is now internalized as the individual effects self-contemplation in contrast to the external reality conditions. The individual can now understand the issues pertaining to the past, individually and collectively, and, in so doing, understand what experiences, methods, or skills to use to foster the development of the new evolutionary purpose in an individual way.

7. **72° — Quintile:** The process of creative transformation through the individualization of the new evolutionary purpose. The meaning becomes highly individualistic and specific. The new individualized purpose is nearly ready for externalized action although it is still somewhat of a struggle to do so because of the pull of the past.

8. **90° — Square:** The individual meaning of the new evolutionary purpose now must be given a new form to operate through in order

to be fully actualized or established within the individual. This "form" can be anything depending on the specific planet in aspect to Pluto. For example, if Mercury is square Pluto the new form would involve a new system of intellectual organization and the resulting opinions. The new form would evolve through analyzing the intrinsic weaknesses or deficiencies in prevailing intellectual systems or bodies of knowledge. This analysis would produce the new intellectual insights. If the planet were Venus, then new forms of relationship or new ways of being in relationship would have to occur. Creative tension is produced through the process of moving forward with this new form versus the compulsive temptation to slide back into old patterns of behavior. This creative tension also manifests itself because the individual may not know how to actualize or establish the new form. The resulting tension or confrontation is usually seen as the individual against him or herself, the individual against another, the individual against society or all that constitutes the past. From an evolutionary standpoint, this form or new way must be actualized or established before the individual can move on.

9. **102° 50' — Biseptile:** Following the square, the new form required at that evolutionary juncture is once again identified as a special destiny of a highly individualized nature. This aspect produces a remembrance of whatever was identified at the septile aspect. Now the desire is to exteriorize this evolutionary cycle or intention, to create a personal reality that reflects this new purpose.

10. **120° — Trine:** The individual is now full of him or herself, so to speak. The desire and intent of the new evolutionary purpose demands creative actualization. This aspect also produces the potential for conscious awareness of the entire process; the past that has led to this moment. The individual can now create a personal reality that clearly reflects the new evolutionary purpose that began at the conjunction. The individual can easily understand what must be done in order to create this reality.

11. **135° — Sesquiquadrate:** This aspect is the point of greatest individualization in relation to the original purpose. It is now willfully expressed through creative activity and action. As the individual attempts to impose his or her personal will upon the environment in relation to this purpose, challenge comes through the external environment. This challenge begins the process of adjusting the individual purpose to reflect the needs of the environment, and the needs of others. This challenge can produce negative results because the individual may refuse to adapt or to adjust his or her newly won and realized purpose — the new evolutionary intent. If

so, then the individual is thrown back upon the past, creating confusion as to how to establish the personal reality and purpose within the external environment until the necessary adjustments are made.

12. **144° — Biquintile:** If the 135° aspect was negatively experienced then this aspect will serve to realign the individual with the original evolutionary purpose by relating it to the individualizing process that took place at the quintile aspect. Analysis must now take place, an analysis as to how to link the evolutionary purpose to the needs of others and the environment. This analysis must be done so that the new purpose can serve the needs of the whole. It is through service that the individual gains a deeper meaning as to the nature of his or her evolutionary purpose.

13. **150° — Quincunx or Inconjunct:** This aspect brings clarification or confusion through self-analysis of the individual's own self-concept as it is identified with the intent of the original evolutionary purpose. The individual is aware that there is something "special" to do in relation to the evolutionary intent, yet does not know how to link this purpose through service to the whole or to others. This aspect enforces some form of crisis in order to induce mental analysis of what needs to be adjusted within the individual in order to establish his or her personal reality or purpose within the framework of the social environment. Humility must replace self-inflated or willfull expression.

14. **154° — Triseptile:** This aspect promotes a clarification of the individual's self-concept and evolutionary purpose as it relates to the needs of others, the environment, or the whole. The necessary self-analysis that promoted a purging of self-inflated delusions of grandeur during the inconjunct aspect now evolves into an essential humility that allows the individual to prepare to integrate his or her purpose within the context of the social environment.

Waning Aspects

15. **180° — Opposition:** At this evolutionary bridge or juncture the individual meaning given to the original evolutionary intent must now be given social meaning. It must be related to and shared with others. This aspect produces the necessity to give the original evolutionary purpose a socialized context or framework to operate through so that the individual can continue his or her ongoing evolutionary development initiated at the conjunction. In order to do so, the individual must enter into social relationships in order to learn that he or she is an equal among others. The individual must learn to listen to others through relationships in order to evaluate

his or her own individuality, and must learn how to relate or apply the evolutionary purpose in a way that is needed by others.

The opposition produces a potential confrontation of will and clashing desires because the individual may feel that their personal power or sense of personal identity is being absorbed and lost through the necessity to interact with others. There may be a sense of losing control of the personal destiny, and the power to shape that destiny from a strictly egocentric point of view. By necessarily having to develop a socialized or more expanded awareness of consciousness, to be pulled out of a self-oriented narcissistic world, the individual may choose to resist this necessity: to return to the past. In reacting negatively, the individual may attempt to dominate others through strength of will, to shove the personal purpose down the throats of others in order to feel powerful and secure.

Until the person succeeds in linking his or her individual purpose to a social need, and becomes an equal socialized being, he or she will remain at this evolutionary gate. The opposing force of the opposition is contained in the ideas, values, beliefs, and needs of others as contrasted with the individual's own needs, ideas, beliefs, and so forth. The opposing force is also reflected in the dual desires inherent in the Soul, which translates into personal will confronting higher will in this context. The desire to return to the Source manifests itself as the need to move onwards, and the desire to maintain separateness manifests itself as the need to remain where one already is for security reasons.

16. **206° — Triseptile:** The original purpose and individual meaning was given social meaning at the opposition. Now it is ready to cooperate with a social or collective need in a realized state.

17. **210° — Quincunx or Inconjunct:** The new social evolutionary purpose or intent serves to clarify the individual's concept or awareness of personal and social limitations; of what he or she can and cannot do, of what is required of the individual by others in order that the social purpose may be expressed. If these limits are transgressed, then intense emotional confrontations will occur in order to reinforce this lesson in awareness. The waxing quincunx induced personal humility. The waning quincunx will induce social humility and purification.

18. **216° — Biquintile:** The socialized evolutionary purpose is now further refined through the individual's awareness of his or her special capacities, abilities, or capabilities as contrasted or evaluated against the capabilities, abilities, or capacities of others.

19. **225° — Sesquiquadrate:** A new crisis now occurs because the

individual must learn everything there is to learn about culture or societal traditions, customs, norms, rules laws, regulations, and taboos before the socialized evolutionary purpose can be established with the society. The crisis occurs because the individual is ready to disseminate the purpose, to apply and establish it, but must learn how to establish it on society's terms.

20. **240° — Trine:** The evolutionary process of expansion and refinement of the abstract and social mind. The individual now has the power and ability to understand how the society works, and on that basis will know how to institute and establish his or her own individualized social purpose within it. The individual can now shape and disseminate the purpose because others (society) will not feel threatened or unnecessarily challenged by the individual.

21. **270° — Square:** Crisis in consciousness. The individual, having learned all there was to learn about society and culture, and having become a socialized being in that context, now experiences a process of repolarization of the consciousness. This crisis revolves around the issues of the past, from where he or she has just been, and the future. The evolutionary intent is to begin a preliminary rebuilding of new foundations of awareness and knowledge that embraces the universal, the timeless, and the absolute. The old patterns, beliefs, time-based cultural truths, and values will no longer serve as the map through which the individual creates social and personal meaning. What to believe, what to think, and how to relate become pressing issues as the individual begins the journey of deconditioning him or herself.

22. **288° — Quintile:** The evolutionary intent is to transform consciousness in relation to individual and social identity. The individual must learn to relate inwardly in an entirely different way because he or she is more than an egocentrically identified or socialized person. This reorienting process is leading the individual into visions and awareness of his or her universal or timeless self, and the relationship to the Ultimate Other. The individual must now learn to see him or herself in relation to the cosmos. There is increasing knowledge or realization of the individual's cosmic role in relation to the functions or duties performed in this life.

23. **300° — Sextile:** The new creative identity of the individual in relation to the cosmic whole is now given productive purpose and understanding to actualize the individual's universal/social role in this life. The transition between the past and the future can be easily made, or easily resisted as this evolutionary juncture.

24. **308°25' — Septile:** Individual action is taken with respect to the perceived universal/social purpose and role by relating this pur-

pose or role to some "special destiny." The potential for misapplication or misidentification of this role and purpose exists; the potential for delusion. However, at key times in life this aspect will produce circumstances or situations to set the individual straight if he or she is in error or confusion. If there is uncertainty about the role or purpose, then circumstances or situations will arise to teach or remind the person of that purpose.

25. **315° — Semi-square:** A new evolutionary crisis emerges as the individual begins to accelerate the mutation between the past, and all that constitutes the personal and cultural past, and the future: the unknown, the unconditioned, the timeless, and the absolute. This crisis is also based upon a conflict or collison of desires and needs. On the one hand, the person wants to withdraw in order to internalize the consciousness so that he or she can become aware of the new seeds or impulses relative to a new evolutionary cycle of development. There is a need for self-contemplation and experimentation with new forms, thoughts, and experiences that reflect these new seeds or impulses; trial and error. These experimentations may create real conflict and confusion. The individual may tire of this condition and attempt to recover the past. If so, eventual disintegration will occur. The individual needs to form new kinds of social relationships with others upon the same evolutionary path. On the other hand, the individual is required to fulfill his or her social duties and obligations, producing conflict between withdrawal and experimentation and fulfillment of the individual's duties and obligations. The key is to do both, and to follow these contrasting rhythms as best as possible when they occur.

26. **320° — Novile:** The seeds of the new evolutionary cycle begin an active gestation process. Like lightbulbs turned on in a dark room, the individual 'magically' becomes aware (revelation) of new thoughts, perceptions, and realizations as to the nature of the new evolutionary cycle and purpose to come. This new cycle is based upon the cumulative effect of actions in this and other lives. Frustration may result from an awareness of the need to complete the karmic and evolutionary intent of this life. Peering over the precipice, the person is ready to jump. Negatively, the individual can become confused and may attempt to retreat into the past as these revelatory thoughts now threaten his or her existing reality and security.

27. **330° — Semi-sextile:** The new evolutionary cycle begins to become clear in the form of complete conceptions and ideas which the individual may attempt to establish in this life. In other words, the individual may attempt to formulate him or herself around timeless values and beliefs in the context of his or her culture; time and

space and all that which is temporal by nature. Considered strange and different by others who do not understand or comprehend what the individual is attempting to do or establish, he or she is given a test to fulfill because of this challenge: to remain committed and centered upon the vision of the new, or the timeless. The whole evolutionary cycle begun way back at the original conjunction is now rapidly dissolving. Some individuals will experience this as a sense of meaninglessness and emptiness, and will manifest a diffuse or undefined personal identity or purpose in relation to the planet aspected by Pluto. The key is to let go of the past in relation to function and orientation, and to allow new patterns, ideas, and impulses to enter the consciousness of the individual of their own accord. Approached in this way, these new thoughts, ideas, and impulses become the switch that illuminates the path to the individual's future.

28. **360° — Conjunction:** An evolutionary cycle has been completed in the personal journey. From the point of the waning semi-square through the conjunction, the process of self-loss or self as a cosmic wholeness (totally integrated selfhood in a universal context) was initiated. Any planet found with Pluto in this condition within the birthchart, from semi-square through the conjunction, has completed, or is completing, an entire evolutionary cycle. A culmination has occurred. The individual will never again experience those planetary functions in the way that they were previously experienced. A totally new evolutionary cycle is about to begin. Planets forming this type of conjunction to Pluto becomes the potential vehicles through which the universal, the timeless, or the Source can be consciously experienced or sensed. Conversely, they can serve as the vehicles through which the individual experiences confusion, disassociation, alienation, and discontentment in order to learn about the nature of personal delusions, dreams and illusions.

We can now discuss Pluto's aspects to specific planets. Remember that all the planets that Pluto aspects must be reformulated, redefined, and metamorphosed relative to preexisting limitations from the past. This evolutionary process has been occurring before. The type of aspect, and the degrees of orb with respect to an absolute aspect, will correlate to how "new" or "old" the process is, and how it has been working before (the type of aspect). In addition, the different ways that this evolutionary process was or will be instigated were described in Chapter One under the section on the Four Ways Pluto Affects Evolution In Our Lives. It may be useful for you to go back and reread this description.

PLUTO IN ASPECT TO THE SUN

In general, when Pluto aspects the Sun there has been and will be an intensified emphasis to creatively develop a special purpose in life. This aspect promotes an awareness within the individual of the power to create a personal reality that reflects this special purpose and reason for being. Thus, the individual has been learning to take control of his or her own destiny, and to shape or create it through strength of will. If the results of this ongoing process have been positive, the individual can creatively transform any activity or area of life to which he or she applies the will and purpose. For this process to be positive, the person must learn how to link the individual purpose to a socially relevant need. If the individual does not learn how to do so, then the possibility arises of remaining in a narcissiastic vacuum of personal creativity that is not allowed to be meaningfully expressed to others.

The evolutionary process has also been teaching these individuals about the limits of personal power — of what they can and cannot do. Even though the lesson has been to teach these individuals how to take control of their lives, and to create their destiny with the strength of their wills, the destiny or purpose itself has some limitation. This limitation is linked to or based upon the actual role that these individuals are des-tined to play within a societal or cultural context. They must learn how to accept this limitation, yet to create their reality, identity, purpose and des-tiny within it.

Often this aspect will promote an active identification with larger-than-life figures who symbolize social power, or who have transcended society altogether. As an example, let's say a person has a serious interest in psychology and wishes to become a psychologist. Com-monly, the individual would identify with, and form a vicarious relationship to, exalted figures within the field of psychology: Jung, Freud, Skinner and so forth. By identifying with the exalted figure the individual would, by osmosis, extract the essence of power of that exalted figure into him or herself: the ideas, principles, methods, techniques, beliefs, and so on. In so doing, the individual would reformulate what was extracted in a uni-que, creative, and personal way.

Pluto aspecting the Sun promotes a cyclic regeneration, renewal, or metamorphosis of the individual's personal creativity and purpose throughout life. Ever greater dimensions of the individual are cyclically revealed. Herein lies a source of potential frustration and limitation. Even though the individual may have a glimpse or even a total aware-ness of what is possible for him or her to achieve and become, the actualization of those potentials must occur at certain stages in life — and only at those stages.

In some cases, depending on the evolutionary and karmic signature, these individuals can attract or come through a parent, commonly the father (or the parent who wields the most authority), who dominates the development of the individual. In other words, the parent attempts to shape the individual's purpose and identity with the strength of his or her will. The individual may feel intimidated in varying degrees of intensity. This intimidation may influence the individual to play out his or her purpose and destiny just as the parent did, and conform to the parent's desires. Such parental dominance tends to extinguish the light of the individual's own purpose and creativity. Yet, because the evolutionary need has been and is for the individual to develop his or her own creative purpose and destiny, this situation will usually promote a "scene" or confrontation at some key developmental point. This confrontation normally leads to a reformulation or metamorphosis of the relationship: either the parent is left in the dust, or the parent changes in such a way as to accommodate and encourage the active development of the individual's own unique identity and purpose.

In other cases, depending upon the evolutionary and karmic signature, these individuals can attract or come through a parent who totally supports and encourages the active development of the individual's own unique and creative identity and purpose. Such a parent will teach that life and reality is what you make of it; you can do anything that you want if you commit yourself to whatever it is that you want to become. In some cases, the individual can attract a parent who fulfills this role, and another parent who fulfills the first role. The specific karmic requirements and needs must be discerned to understand why this has happened, and what to do about it.

In general, then, those with Pluto aspecting the Sun have something very special to accomplish. This purpose can lead them into a life of relative fame, recognition, adulation, and acclaim relative to what their special purpose is. In certain karmic situations generally associated with an abuse of power, or refusal to acknowledge the limits of that power, this acknowledgement and acclaim will not occur. In this case, the individual will live a life of exposure to the achievements of others while being reduced to relative insignificance by society or others. The enforced limitation must be used to develop an objective awareness as to why this situation has occurred. The individual can then grow in such a way as to be free of this condition in this life, or be freed in the next life experience or incarnation.

It is important to remember that this aspect promotes a compulsive need to eliminate all conditions or dynamics that prevent the uncovering or realization of the individual's personal core of power and individuality. Because the person has been learning to creatively express or project this personal power and purpose, there can be an unconscious desire to be acknowledged as special, powerful, or important and thus create a danger of getting hung up on self-glorification. On the negative side, the means to achieve power can be quite underhanded, manipulative or ruthless, even criminal in certain cases. In a positive expression, the person would not permit him or herself to fall prey to these means of achieving his or her special purpose or destiny. The individual would realize or understand the legitimate means to establish and actualize his or her self, and would challenge or attack those that attained power and recognition through negative or illegitimate means.

Many people with Sun/Pluto aspects are compulsively drawn to controlling or manipulating the individual expression of other people, especially those with whom they are in close relationship. In this negative way, they are continually trying to reformulate the behavior of another according to the standards of conduct that they feel is right. Of course, this attitude normally promotes confrontation and emotionally tense scenes, and separations often occur. The resulting emotional shock will leave the individual in varying degrees of devastation. On the brighter side, this experience will promote an awareness of the dynamics operating within the individual that created the problem.

Conversely, some individuals will seek out relationships or situations in which they become absorbed or merged with the power and will of another. In this negative way, they unconsciously desire and attempt to become powerful and meaningful through a vicarious association with another's power and purpose. In the end, such an association will degenerate until there is a parting of the ways. This will occur through necessary confrontation so that these people learn to create their own purpose and destiny based on their own self-actualized efforts.

Positively, individuals with the Sun/Pluto aspect will not only actualize their own power and creative purpose, but also encourage others to actualize themselves along the lines that are most natural to their own development and evolutionary capacities. They will perpetually act in this way, and will not tolerate others who do not. The very symbol of their life can serve as an example for others to develop the courage to be free of the unnecessary limitations in their own lives.

PLUTO IN ASPECT TO THE MOON

The prior evolutionary intent of this aspect was to eliminate all external dependencies in order to induce the lesson of internal security. In addition, this aspect has served to reformulate or transform the individual's instinctive emotional reactions to any internal or external circumstance. In both ways, the individual has been learning to transform the self-image; how he or she sees, identifies, and relates to him or herself.

Often this aspect has promoted emotionally difficult experiences with female or mother-type figures. Commonly, the female figures or mother-types have been extremely dominant, control-oriented people with an intense and rigid standard of conduct or behavior that was projected upon the person with this aspect. When the individual failed or refused to conform to these compulsive dictates and expectations, emotional blows or attacks were delivered upon the individual. Or, the female or mother-type withheld emotional expression in icy silence; the vibratory projection of disapproval, letdown, or hurt permeating their auric field. In either case, the individual was thrown inward to induce the necessary evolutionary lesson.

As a result, these individuals commonly have displaced and unresolved emotional needs and problems that are projected via expectations upon others. When others do not meet these displaced emotional needs, nor solve the problems, the projection of emotional rage or anger will result. This experience, of course, enforces a confrontation with others through which the individual is thrown inward. In another common reaction to this process, the individual will withhold emotional relations with another when his or her deep-seated needs are not being met. The intensity and degree of these experiences can be linked to how "new" or "old" this evolutionary process is, and with the type of aspect formed with Pluto. For example, a client of mine who had Pluto in Leo retrograde conjunct Saturn retrograde in the Ninth House inconjunct a Pisces Moon in the Fourth House experienced the separation of his parents when he was six years old. He was put into an orphanage at that time. The parents subsequently got back together and brought him home. When he was ten he overheard his parents discussing the fact that they did not like him and wished that he had not been born. Upon hearing this conversation he left home and walked twenty-five miles to his godmother's house. The godmother loved him, but the parents took him back. Thus, he was forced to live in an environment wherein he was not accepted, loved, or understood. Because the inconjunct aspect to Pluto was four degrees after the exact aspect the psychological and emotional effect of this rejection was not severe. He actually had memories of being through these kinds of experiences in other lives, and had more or less learned to minimize his emotional expectations and dependencies. The evolutionary need to come through this kind of parental environment was to induce the final lessons to this end.

Conversely, some of these individuals will come through or be exposed to female or mother-type figures who encourage emotional self-reliance and inner security. These female or mother-types will have the ability to penetrate and understand the emotional dynamics of these individuals and, in so doing, help them to unravel the basis and causes of their emotional winds, moods and feelings.

The evolutionary intent of this aspect demonstrates that these individuals have been learning how to reformulate their emotional conduct and self-image. Emotions can be compulsively expressed when Pluto aspects the Moon. The person may emotionally and intellectually resolve to operate in a different manner, yet, much to their own horror, find themselves manifesting an old compulsive pattern in emotional behavior. This situation promotes self-rage, anger and hatred on a cyclic basis because the individual feels powerless to change the pattern even when the desire and intent is to do so. In the worst scenarios, this dynmic can promote thoughts of suicide, or a subconscious death wish as the individual seeks to escape the pain that characterizes his or her emotional life.

These individuals have been learning how to shed light upon their inner emotional dynamics: how and why they work the way they do, and for what reasons. As a result, they will be intensely focused upon the emotional dynamics, intentions and motivations of others in order to learn where they are coming from. Based on their own emotional compression and intensity, they commonly have an innate ability to understand the emotional psychology of other people if the aspect between Pluto and the Moon is relatively "old." Some will compulsively attempt to reformulate the emotional expression and self-image of others, just as the female or mother-type attempted to do in the individual's own life. In other words, they duplicate the behavior of the mother, or key female types in their life. Most people will not tolerate such emotionally overbearing behavior for long. Many relationships approached in this way terminate in confrontations and emotional shocks. When others do not conform or meet the emotional demands and needs of these types, they can become very cruel, vindictive, spiteful and jealous. Some will become physically violent when inner turmoil and rage breaks loose from the tight reins of control that the individual normally maintains. On the other hand, some Moon/Pluto individuals will attempt to motivate others through helping them unravel their emotional dynamics in a non-manipulative way. They will encourage emotional self-reliance and inner security, not dependency upon them. Such individuals will have experienced female or mother-types who did this for them, or they have evolved into this condition through the emotional school of hard knocks.

Once the lesson of internal security have been consciously sought or initiated as a result of this evolutionary process, all the negative emotional behavior progressively turns positive. Progressively these individuals learn to fulfill their own emotional needs and, in so doing, they free themselves to operate within relationships and life in a non-demanding, self-secure way. Now they can display patience and tolerance where there once was intolerance and lack of patience. Now they can encourage inner emotional growth, health, and independence in others. Where the individual once repelled others, now he or she magnetically attracts others because they recognize his or her natural healing abilities.

The Moon represents the psychological function of the ego. Those with Pluto in aspect to the Moon have been learning how to focus the power of their Souls through their conscious ego. As a result, they can be incredibly self-determined and singleminded in their pursuit of an objective — whatever it may be. They can harness the penetrating quality of Pluto to uncover the "bottom line" of how anything is structured or works. In so doing, these individuals will instinctively know how to manipulate or use the means necessary to accomplish whatever goal or task they desire to fulfill. The means can be ruthless or manipulative through negative expression, or extremely above-board when positively expressed. These individuals commonly have a deep and penetrating gaze because of the Moon's symbolic correlation to the retina in the eyes. The force of their Souls is reflected through their eyes as they attempt to penetrate to the core of others to see where another is coming from.

Because the Moon also constitutes our personal environment, Pluto/Moon individuals can tend to dominate their environments through the sheer intensity of their energy field that is instinctively expressed through their egos. This dominance can be expressed through utter silence as well as through communication or action. In either case, their "presence" is felt by others. Often they suffer or experience misinterpretation or misidentification of their intentions. This occurs because the intensity of their emotional body and energy field can create an enigmatic effect compounded by the natural secretiveness and mystique of Pluto. This kind of environmental experience can be very frustrating for these individuals. From an evolutionary point of view, this kind of environmental challenge is meant to draw the person out of him or herself to develop an awareness of what he or she is actually feeling or thinking. This challenge also promotes an inner examination that allows for the individual to check out his or her motivations, intentions, or the basis of what is occurring from within that creates this kind of challenge. An inner examination must occur because these individuals cyclically "shut down" on an emotional basis. The danger in shutting down is one of emotional and psychological implosion wherein all perspective is lost. If the individual is in a shutdown cycle, then the challenge from the environment will attempt to pull the person out of it. If the individual is in an animated state, then the challenge from the environment will attempt to question the basis of where the individual is coming from. Often this challenge is met with resistance as the individual refuses to reveal what is happening inside. Or the individual may refuse to accept or acknowledge the correctness of another's perception or insight when challenged in this way. This issue is very important to understand for those who interact with these individuals. It is important because by nature these people *need* to cyclically shut down in order to replenish their emotional batteries. They also need to shut down because they are cyclically consumed by the unconscious force of their Souls. This occurs in order to induce knowledge or perspective, or to release a new seed thought, feeling, or emotion from the Soul to the conscious ego at key times in the evolutionary journey. Thus, the balance between necessary shutdown time and animated time is critical. Too much of either will promote distortion and the loss of a center of gravity for these people. When challenges occur from the environment with respect to each cycle, these individuals should learn to pay attention to them because they are normally "signals" that an extreme is being reached. If the individual is in an animated state, then these challenges will reverse the flow because of the animated response that they demand. Paying attention to these environmental signals can promote a state of emotional and psychological balance in these individuals. Resisting these signals can promote imbalance.

Moon/Pluto individuals are also emotionally sensitive to an intense degree. Conditions in their environment have to be "just right." If they are not, then they may become very upset, and attempt to make it right. The behavior of others must conform to their standards of conduct. Of course, this expectation promotes confrontations through which the individual learns to change his or her instinctive emotional responses to anything that does not reflect the standard of right conduct. The obvious limitation of their rigid and fixed standards must be confronted so that a metamorphosis can occur. The individual then learns that other standards of conduct, values, needs and beliefs are just as powerful, legitimate and relevant as his or her own.

This aspect also commonly promotes intense emotional and sexual needs. A cyclic buildup of emotional energy demands release, projection, or expression. This release can take place through sexual activity. The sexual needs projected upon others can be quite intense and demanding if the individual has not learned the lesson of emotional self-sufficiency, which leads into sexual self-sufficiency (masturbation). Sexual release promotes emotional stability and healing for these individuals. Normally the sexual release (orgasm) is utterly consuming and total. For awhile the individual can be at ease and in an emotionally relaxed state as a result.

It is a good idea for Moon/Pluto individuals to commit themselves to an intense and concentrated program or system that allows for a personal transformation or repolarization of their instinctive emotional reaction patterns so that they are able to adjust to life in a more objective and open manner. This is a good idea because many of the emotional memories that are held in the unconscious are painful and difficult. These memories dictate their emotional reactions in a compulsive and "irrational" manner. These individuals can become consumed by their own moods, feelings and emotions. These moods, from the darkest to the brightest, and all shades in between, originate in the distant and unconscious memories. The need to gain perspective and control of these states is critical. This perspective and control can occur through a sustained commitment to some program or system that allows for the development of objective of emotional awareness.

PLUTO IN ASPECT TO MERCURY

In general, Pluto/Mercury contacts have promoted necessary intellectual confrontations with others so that rigid and fixed ideas could be reformulated, transformed and expanded. Individuals with this aspect have linked much of their emotional security and stability to their ability and need to intellectually organize reality in a way that reflects their ultimate beliefs and values. Because emotional security is connected to an intense power of intellectual organization these people tend to defensively and compulsively maintain their ideas. To change an idea, to consider something in a different way, is to risk emotional insecurity and a sense of powerlessness. The evolutionary need has been, and is for, confrontation. Anything that is relatively fixed will stagnate and prevent further growth. Confrontation forces these individuals to rethink or reexamine their most cherished opinions — whether they want to or not.

Commonly, these individuals have intense powers of mental concentration. This ability allows them to penetrate to the bottom line of anything they turn their mental gaze upon. By analyzing and focusing upon the bottom lines, they can determine the inherent structure, the essence, of what they are examining. Yet, the interpretation of that which is being examined will always reflect the individual's preexisting beliefs. This belief system creates a filter. That which is being examined, studied, or analyzed will be rejected if it does not in some way reflect what the individual already thinks is true or false. In other words, the individual will not allow the information to penetrate. To do so would be to challenge preexisting beliefs, and undermine the sense of security which is bound by the individual's opinions and ideas as to the nature of things. On the other hand, if the individual desires to learn more, and if the experience also supports preexisting beliefs in some way, then he or she will methodically and thoroughly take in that information. It will become his or her knowledge through the osmosis effect of Pluto.

It is very important to examine the actual nature and condition of

Mercury itself. What house and sign is it in? What kind of aspect exists between Mercury and Pluto? Mercury by nature is curious about everything. Mercury wants to experience intellectually many things, and it wants to communicate its knowledge to others. It wants to enter into intellectual discourse with others. Mercury wants to learn. And it wants to organize intellectually that which it is experiencing in a logical way — to make connections.

Two very extreme effects can occur when Pluto aspects Mercury. On the one hand, the individual may totally preclude intellectually experiencing anything new that does not reflect what he or she already feels to be worthy or useful to preexisting interests in life. This extreme would promote a narrowness of intellectual focus. With respect to com-munication, the individual would not communicate to anyone unless there was a reason or purpose to do so. There would be no unnecessary communication or idle chit-chat. And when this individual did com-municate, it would be to the point; no mincing of words. The tone of communication would be emotionally powerful, almost hypnotic in effect, as the strength of the Soul would be reflected in the individual's intellectual convictions and opinions. On the other hand, the individual would be intellectually open to everything. Intense mental curiosity would drive the person ever onward in a search for knowledge. This type would feel compulsively drawn to communicate with everyone, to study everything in the quest for information and knowledge. Whatever the individual believed in or had an opinion about, would be powerfully com-municated with emotional intensity and conviction. Between these two extremes, of course, are many shades or combinations of expression.

So it is important not only to understand the total nature of the individual, but more specifically the condition of Mercury by house and sign locality, planetary aspects, and the type of aspect to Pluto. For example, Mercury in Libra in the Third House, sextile Pluto in the First House would incline the individual to be the "open" type. Conversely, Mercury in Taurus in the Eighth House, square Pluto in the Fifth House, would produce a "closed" type who is destined to experience two types of confrontation:

1. They will experience external confrontations with other individuals that test, challenge, or question their intellectual organization and opinions. The narrowness and subjectivity of their specific mental focus is the basic problem. The problem is one of limitation that promotes black and white points of view. Every experience that the individual encoun-ters will be forced into these black and white categories. The particular point of view that the individual holds is not necessarily wrong. Yet, it is limited. From an evolutionary point of view, confrontations must occur. The effect is to produce a crack or opening in the rigid intellectual

categories. New information can then come in to reform or expand the intellectual information base, which in turn can create new perspectives.

The resistance in these individuals to these confrontations can be quite intense because their emotional stability and security is at stake. This resistance can create the subconscious effect of the person not listening to others and repelling even the most convincing argument delivered by another. This resistance usually occurs at the time of the confrontation. After the fact, the individual may reflect upon what has happened and in his or her own time may choose to consider or incorporate some new information or point of view into the intellectual data base. This will occur if the individual can perceive the usefulness or legitimacy of doing so. When these individuals repel another's point of view in order to defend their own, they will focus on the weakest possible link in any argument or intellectual system of organization and use that weak link as the basis of their rejection.

2. The black and white intellectual system can also implode from within. The person will cyclically experience times in which his or her way of intellectually understanding him or herself, others, and life does not work. Loss of perspective occurs. In these cycles of intellectual implosion the person will desperately reach out for new information or ideas that can create or lead to a new perspective for understanding the basis or nature of whatever problem is at hand that created the necessity of this implosion.

With the open type of individual, the confrontations will occur with respect to their inability to commit themselves to any specific point of view beyond a certain duration in time. Becoming lost to their own internal revolving door of mental perspectives, these individuals can see the relevancy or legitimacy of all points of view. This situation promotes a mental crisis because the evolutionary need is to commit to a specific and particular system of knowledge that allows all other points of view to be referred, and therefore emotionally and intellectually integrated into their total consciousness. The mental crisis produced through this evolutionary need creates a confrontation as to what system to commit to. Experiencing the power and intensity of so many others' points of view, reading parts of this book and that, taking this seminar and that class, only compounds the problem. Loss of perspective occurs as the individual desperately races back and forth from one idea to another in order to explain the nature of the problem. Or, engaging in a conversation, the individual will compulsively take the other side of any argument or exchange in order to prove the usefulness of another point of view.

What to believe, what to think, and how to think about it become major dilemmas. The evolutionary solution in this situation is to choose any system with which the individual naturally resonates on an emotional level. By choosing one system, clarity of thought and perspective will follow, which in turn will promote greater emotional stability and security. The challenge, however, is not to defend this system as the only "right" way, as that attitude would create a "closed" type of individual. The challenge is simply to commit to and absorb one system and to use it to generate an intellectual and emotional cohesion. The person now should adopt the attitude that their own system works for them, and other systems or points of view are fine for other people. Rather than defending or arguing, the person can now use his or her own intellectual power base as a foundation for sharing ideas. This inner foundation would now allow for development of a healthy and valid discrimination so that the individual does not fall prey to every passing intellectual statement or idea that piques the curiosity.

In almost all cases with Pluto in aspect to Mercury, there has been, and will be, a mental need to explore, experience, or investigate areas of life that are considered taboo by society or the parental environment. These individuals reason that the taboo may give a piece of knowledge or information necessary in order to understand the mysteries of life. Taboos, again, imply limitations. The evolutionary need is to grow beyond that which promotes limitations.

PLUTO IN ASPECT TO VENUS

In general, when Pluto is in aspect to Venus, there has been and will be a long-term evolutionary conflict based on the dual nature and requirements of Venus. As an inherent archetype or dynamic, Venus demands that before we can successfully be involved in relationships of any kind, we must learn how to relate to ourselves, love ourselves, and to supply and answer our own needs in a self-reliant and self-sufficient way. It is only when we learn how to do this that we can enter into a relationship successfully. Actualizing this dynamic and need will allow us to be in a relationship with a minimum of projected expectations and needs that we expect our partner to meet or satisfy for us. If we cannot love and relate to ourselves, how can we really love and relate to another?

Pluto in aspect to Venus demands that we learn to merge ourselves with another. Through this experience a transformation of personal limitations occurs for both individuals. Because the evolutionary demand has been to merge with another, most individuals with Pluto/Venus aspects have not learned how to relate to themselves independent of relationships. They have not learned to be self-reliant, self-sustaining individuals who are able to identify and answer their own needs outside of a relationship. In most cases, they will feel that they cannot be fulfilled or completed except in relationship to another. Herein lies the long-term evolutionary dilemma.

Because of the Pluto connection, the need to explore and discover the deeper meanings of life, these individuals have commonly been drawn in a hypnotic-like way to other people who symbolized something that they needed. This dynamic implies that whenever this kind of attraction occurs, the individual does not already possess that which is needed or desired. These individuals can be deluded into feeling that the meaning that they are seeking is embodied in another, and is not within themselves. In order to possess or secure a partner who apparently has what they feel they need, these individuals can emotionally manipulate another in order to become involved. The forms and strategies that this manipulation takes are unlimited: from covert and obtuse, to overt and direct. Bear in mind that others can approach these individuals in the same way, for the same reasons, and with the same strategies. Again, external reality is a metaphor for our inner reality.

Such relationships have created many psychological, emotional and karmic problems including intense and unrealistic expectations projected upon another, or projected upon the Venus/Pluto person by another. The giving of love and the fulfilling of another's needs has been conditioned by having one's own needs met. If they are not met, the withholding of giving or love can occur. This obvious confrontation of whose needs are going to be met or not met has promoted many difficult emotional scenes. In many cases, these scenes have finally led to a termination of the relationship. Often, these partings were very difficult.

Emotional, psychological, or physical abuse is not all that uncommon an experience for these individuals over many lifetimes. The memories of these experiences are held in the unconscious memory banks. Thus, many of these individuals will come into this life with a natural suspicion of others (where are they really coming from?, what are their motives?). In addition, there is commonly a fear of having the rug pulled out from under their feet, giving these individuals the appearance of emotional aloofness, coldness, or an inability to commit totally in a relationship. Karmically, especially when stressful aspects are involved, these individuals will draw, or be drawn to, others with whom they have been involved before in previous lives. The theme in these relationships is almost always to resolve an issue that was not resolved previously.

Another common problem for many of these individuals is one wherein they have only been able to sustain a relationship for the duration of the need that created the relationship. As a result, there have been many cases wherein one partner may feel there is no longer a reason to be together, while the other wants the relationship to continue because he or she did not feel it was over. Individuals with this aspect have been in both positions: they have been left and have been the one leaving. The basis of this problem, again, is seeking personal fulfillment and meaning outside of one's self.

Even when in a primary relationship, those with Pluto/Venus aspects may experience intense hypnotic attractions to others. When this occurs, the attraction will always be based upon some need that is not currently being fulfilled in the existing relationship. At this point, a problem has been created. As a general principle, the need of Pluto/Venus aspects is to commit to one person. How the individual reacts to this problem determines the karma that will follow. If the individual left the existing relationship to follow the hypnotic attraction, karma is set in motion with the first partner that must be fulfilled at some other time. If the individual did not follow the hypnotic attraction, then the unresolved desires associated with that person must be fulfilled at some later time.

Another problem with this approach to relationships is one wherein one person will wield more power within the relationship than the other. If the individual follows an attraction primarily based upon what he or she needs and what the other can give, then the other person is automatically placed in a position of power and control. This dynamic obviously creates inequality and potential confrontations as the other person compulsively manipulates the Venus/Pluto individual to maintain that position. On the other hand, if another is attracted to the Venus/Pluto person because of their projected needs, the Venus/Pluto individual is in the position of control. Again, confrontations occur for the same reasons. The Venus/Pluto person has played both roles in prior lifetimes.

Almost all people with this aspect have needed to confront or challenge the socially defined way of relating to and being involved with others. They have been desiring and needing to eliminate the "taboos" of how to be in relationships. They usually do not challenge the taboos with direct assault upon existing values, but hold these needs and desires within themselves. Thus, the desires and needs attract others who seek the same experience. Within the privacy of their own relationship these individuals explore a plurality of possibilities that each feel naturally drawn to investigate on an experiential basis.

Many of these individuals have a need to control or be controlled in varying degrees of intensity, and commonly desire to possess another because of their emotional and stability needs. Again, this situation will occur because these individuals are looking to another to fulfill their needs, or another is looking to them for the same reasons. This form of control and emotional limitations can only promote stagnation for both people. At some point a critical mass situation will occur wherein the enforced limitations of this form of relationship will explode or implode.

Confrontations or separations will result with the karmic consequences mentioned above. The degree of control and possession is usually dependent on how much the individuals really want to be with one another. The degree of want determines the degree of need. The degree of need determines the level of control.

Negatively, these individuals can be very jealous, possessive, vindictive, cruel and mean when they feel their partner has done something to threaten or undermine the relationship, or has done something to abuse or hurt them, or has failed to acknowledge their needs. All too often, the Pluto/Venus person will withhold their affection, love, or feelings from another for fear of being let down or abused, or as a means of controlling the situation through the intensity of their self-imposed isolation. And, of course, they can draw others to them who respond in the same way.

All individuals with the Pluto connection to Venus need to know why they feel as they do, and why another feels the way that they do. They need to understand why they need what they need, and why another needs what they need. Over time, this aspect has produced a natural capacity to get inside other people to know how and why they need what they need. This ability creates a naturally intense magnetism that can attract others who feel that these individuals can understand and fulfill their needs. Conversely, the Pluto/Venus person can be attracted to others who are also intensely magnetic for the same reasons. In general, these individuals are attracted to intense people who are dealing with the deeper issues of life. They are not attracted to superficial or "ordinary" people. They have a naturally psychological approach to relationships. They are always listening or trying to understand the deeper meaning behind the words or actions of others, and others respond to these individuals in the same way.

Pluto in aspect to Venus produces individuals who are very passionate and sensual. These individuals enjoy and require touch and the warmth of bodily exchange. In fact, their touch alone will act as a barometer that reflects their feelings about themselves and how they are feeling toward those they feel inclined to touch in an intimate way.

All the potential problems and karmic situations can be instantly or progressively metamorphosed once these individuals learn to relate to themselves outside of a relationship and to identify and supply their own needs. Essential self-relatedness and self-love will eliminate the need to project unrealistic needs and expectations onto the partner. In addition, these individuals will learn how to encourage and support independence and self-reliance in their partners rather than trying to control them. The intent of the confrontations which occur in the relationship is to promote this lesson. Once these people stop looking outside of themselves for the meaning of their lives, they will be able to choose and commit themselves to a partner who reflects their own higher evolutionary and karmic needs. In this way the relationship can thrive and grow. In certain difficult karmic situations, some of these individuals must experience the emotional shock of losing a mate through death in order to enforce this lesson. Even the shock of the death of other close people in their lives can induce this lesson. Once the energies of the Venus/Pluto aspects are metamorphosed, these individuals become not only the closest and most trusted and loyal lovers or friends, but also individuals who support and encourage the needs of others to transform the limitations and blockages of their own lives in order to be free.

PLUTO IN ASPECT TO MARS

Before we discuss the specifics of Pluto aspected to Mars, a general discussion of the connection between these two planets is in order.

In traditional astrology we have learned that Mars is a lower octave of Pluto. A lower octave is a denser expression of a higher vibration. From an evolutionary point of view, Pluto is correlated to the Soul. As explained in Chapter One, the Soul has two archetypal desires: one to separate from the Source of the Soul, and one to return to the Source. The dual desires coexist and interact. The Soul is rooted in our unconscious. How then do most of us know that the Soul exists beyond the testimony of enlightened beings or spiritual teachers and their teachings? We can all know that the Soul exists by the simple fact that we have desires. All of us have myriad desires that dictate what we think we need, which leads to the choices that we make. These choices determine the actions that we take. Actions lead to reactions, which create new actions and so on.

Mars is the lower octave of Pluto and thus correlates to the conscious component in our personalities that instinctively initiates or acts upon the desires originating from the Soul. Mars represents the instinctive urge in all of us to act, to become, to move forward with our lives. Is it not true that we are all in a continual state of becoming at each moment of our lives? To become is to act, to be in motion. And each of us is becoming who we are as unique individuals according to karmic (action/reaction) and evolutionary laws. These laws are determined by our previous, as well as our current, desires. If I kill someone this very moment, I myself may be killed in the very next moment.

Because Mars is conscious and Pluto unconscious most of us feel that our desires originate at a conscious, egocentric level: my ego is in control. The fact is that all desires originate from the Soul, and are transmitted to the individualizing aspect in all of us: Mars. Mars instinctively acts to fulfill the desires originating in the Soul. This dynamic is the very basis of the Indian concept of *Maya* — the illusion of separateness from the universal whole or Source. The classic example of the wave upon the sea will suffice to illustrate this point. If my consciousness is centered in the wave, then from the wave's perspective I am an egocentric individual. If my consciousness is centered in the sea, the source of the wave, then I realize I am but an individualized aspect of the sea: the wave. As the wave must return to the sea, so too must all of us in our evolutionary journey return to the Source. Mars is the wave. Pluto, the Soul, is the source of the wave. Yet Pluto, the Soul, is also a wave of the Ultimate Source; that which created the Soul as well as everything else.

Just as the wave will resist returning to the sea for a time — because of its upward motion and momentum — so too do all of us resist returning to the Source that created us. Mars represents the personal, conscious and subjective will that is centered in our personality. The Soul, because it contains dual desires, simultaneously correlates to the will of the Source and the will to remain separated from that Source. This dynamic is the basis for conflict within an individual, between individuals, and between an individual and the Source. This conflict originates in the Soul and is reflected through our personalities. Our personalities are created by our Souls, relative to the karmic and evolutionary laws set in motion by each of us. The karmic and evolutionary laws are in direct relationship to our prior actions and desires. Again, Pluto transmits its desires to Mars. With respect to the coexisting desires in the Soul, is it not true that each of us will consciously sense that there is "something more" after we have obtained something that we have desired of a separating nature? In a conscious way Mars reflects the unconscious desire to separate and return. Thus we can desire the new lover, position, possession, and so forth, yet still feel that there is more. We are dissatisfied after the temporary glow of actualizing what we have desired of a separating nature. It is this sense of dissatisfaction that consiously reflects the unconscious desire to return to the Source — the ultimate satisfaction.

The point is this: the relationship of Mars to Pluto in the birthchart, the house and sign locality of each, and the aspect or phase linking them, will describe how the individual acts to implement (Mars) the desires originating from the Soul (Pluto). The house and sign locality of Mars represents an area in the birth chart through which many kinds of desires are manifested or transmitted in order to become, at every moment in time, who we are destined to become according to our karmic and evolutionary laws. It should be clear that free will or choice exists in the Soul relative to which desires are acted upon, and which are not. Via transmission to Mars, this phenomenon of free will and choice is also reflected in our conscious personalities.

A simple example will illustrate these ideas. Let's put Pluto in Leo in the ninth House trine Mars in Sagittarius in the Second House. Let's assume the individual is in an individuated evolutionary state. As stated before, an individual with Pluto in the Ninth House will naturally ponder metaphysical, cosmological or philosophical questions concerning the nature of existence in an effort to understand him or herself in this context. The Soul has desired this knowledge before, and will naturally desire to expand on this orientation coming into this life. The evolutionary intent for this life is reflected in the Third House and Aquarius polarity: to learn the relativity of philosophical, cosmological, metaphysical or religious teachings. In other words, that the personal version of the cosmological order (beliefs), and the way to realize the truth (path) of that order, is relative.

There are many paths to the truth. Internal and external philosophical and intellectual confrontations would lead to this lesson. They would promote a necessary objectivity. With Mars in the Second House, the individual would naturally manifest desires (originating from the Ninth House Pluto in Leo) to be free in order to explore whatever experiences were necessary in order to learn about life because of these lessons. Through exposure to many different kinds of experiences the individual could learn not only about larger metaphysical connections, but also the relativity of differing ideas or systems. With Mars in the Second House, the person would naturally desire to be relatively self-contained, self-sufficient and self-contemplating. This individual may naturally desire to study nature, for example, and would contemplate natural laws as revealed in nature. In this way, the individual would understand him or herself in a cosmological and metaphysical way because of his or her relationship to nature and the natural laws therein. The waxing trine between Mars and Pluto would suggest an ease of understanding of this process. It would suggest that the person would synthesize all experiences and thoughts into one belief system of his or her own design that would serve as the basis of his or her understanding about the nature of life.

Mars in the Second House would also reflect the desire to identify personal resources that could be used to sustain or support physical survival. With Pluto in Leo in the Ninth House, a natural gift of teaching could be a possible means of creatively actualizing the individual's own unique purpose and destiny. This evolutionary capacity and desire would manifest itself through the Second House Mars in Sagittarius as a desire to teach as a means of self-support. By developing the teaching capacity, the person would be naturally exposed to the ideas, philosophies and opinions of others. This dynamic would naturally create the evolutionary desires for this life: to learn the relativity of belief systems, the different paths or approaches to truth, and to actualize the creative purpose or destiny by linking it to a socially relevant need. Expansion upon previous development and objectivity would occur as by-products of this evolutionary direction and commitment. Negatively, the Second House Mars would manifest a desire to withdraw again in order to not face the confrontations that threatened the emotional security of the Ninth House Pluto and the deepest preexisting beliefs. The Second House Mars would reflect the transmitted desire of the Ninth House Pluto in this way.

All individuals with Pluto/Mars aspects have been, and will be, learning karmic and evolutionary lessons about the nature of their desires. Ultimately we all face this lesson, but the need is emphasized by these aspects. Mars in aspect to the South or North Nodes would also correlate to this same lesson.

The key to working with the tremendous amount of raw energy and power inherent in the Mars/Pluto aspects is to direct it toward goals that are personally relevant to the individual. By channeling this energy and power, these individuals will be able to effect a consistency and continuity of action in their lives. By harnessing this power through linkage to relevant goals, they can create a dynamic for processing their energies. In other words, self-knowledge and understanding can occur through the activity that the commitment to a goal requires. Until such a commitment is made, many of these individuals will follow one desire after another, each desire taking them to one experience after another. Dissipation of personal energy will occur. As a result these individuals can experience a loss of perspective as to who they are and what they are doing — and for what reasons.

All those with Mars/Pluto aspects have a deep-seated need to transform the limitations of personal reality and individual identity. Personal energy must be channeled and linked to a relevant activity in order for this transformation and evolution to take place in a positive manner. Water passing through the generators in a dam offers an apt analogy. The dam channels water through portals. The water turns the generators, which produce tremendous quantities of electricity. In the same way, channeling energy will allow for personal transformation of preexisting limitations so the Mars/Pluto individual is in a continual state of 'becoming' at every moment in life.

Commonly, these individuals will resent any restrictions or limitations placed upon them by external agents or conditions. They will also resent any restrictions of limitations that they find within themselves. The desire and compelling drive to eliminate all limitations reflects the need to discover who they essentially are at a core level. The house locations of Pluto and Mars, plus their signs and aspects, will suggest or describe how this personal transformation will take place relative to the evolutionary intent and desires that these individuals have. The potential for inner conflict is enormous. As an example, let's say Pluto is in the Fourth House in Cancer and that Mars is in Libra in the Seventh House. The two planets are in a waxing square. On the one hand, this individual would desire to explore a variety of experiences through relationships that were formed on an instinctive basis. In so doing, the individual would progressively discover or realize ever new dimensions or aspects of him or herself.

With Pluto square Mars, many of these instinctive encounters with others could be of a sexual nature or the individual would be tempted to sexualize them. On the other hand, Pluto in Cancer in the Fourth House commonly needs familiarity, continuity, and a security that is linked to home or family environments. Thus, the conflict or clash of desires would revolve around the individual's need to explore new situations and relationships, and the need to maintain the old and the familiar in order to feel secure. Commonly, this pattern would promote a strong-willed partner who instinctively and compulsively tried to control the emotional and individual definition of the Mars/Pluto person. Of course, this situation would promote inner rage and anger as the individual experienced his or her limitations and restrictions in this kind of situation. Yet, because of security needs, the individual may not act upon the conflicting desire and the need to break free from this limitation. In essence, this individual subconsciously attracted (desired) this kind of partner in order to feel protected and cared for. This partner was also subconsciously drawn to enforce the individual's evolutionary conflict. At some point, the individual may erupt out of the relationship once the restriction and control reached a critical mass situation, as would almost inevitably occur because of the nature of the waxing square between Mars and Pluto. The need for "new forms" or new ways of being in a relationship would only increase the evolutionary conflict and energy to shatter this limitation. The clash of wills between the two partners would aggravate the conflict to further compel termination of this situation.

When the lid finally blows, a metamorphosis will occur. The essence of this transformation will be the discovery or realization of the individual's own core, and the ability to stand upon his or her own feet in a self-secure way. In addition, the individual would come to recognize the need for relative equality (Mars in Libra) within a relationship wherein the partner encouraged the freedom and independence to become what the individual must become in a self-determined way. In this way, this kind of relationship would allow for a channeling of personal desire and transformation, versus running around and going through one partner after another. This latter approach would promote an ultimate dissipation of energy and loss of personal perspective. This situation would ultimately enforce the lesson or realization as to the nature or basis of the individual's desires and evolutionary needs, and how the interaction of the conflicting desires and needs created, shaped, or led to the conditions of their reality.

In general, Mars/Pluto aspects promote a tremendous amount of sexual energy. The intense emotional power of Pluto is now linked to the physical nature of Mars. The buildup of emotional/sexual energy must find a release. If it does not, then emotional distortions of all kinds can result. This dynamic has two possible extremes of expression: 1. to be driven and controlled by the sexual desire nature, or 2. to be in control of the sexual desire nature. The stressful aspects tend to promote more extreme levels of emotional/sexual energy than do the nonstressful aspects.

The desire and need is to exhaust or rejuvenate one's self through sexual experiences. Many will experience a personal transformation of limitations through emotional/sexual experiences. There is a compelling need for this kind of experience. Emotional/sexual energy can take many directions within these individuals. The actual direction depends on the individual's overall values and belief systems, and the condition or placements of Mars and Pluto within the birth chart. Some will transmute their sexual energy by focusing upon the fulfillment of a goal or ambition. By using and directing the emotional/sexual energy in this way, the individual becomes consumed by the ambition or goal itself through which a transformation or personal limitations can occur.

A couple I once counseled offer an interesting example of this transmutation process. The man was a rock musician. The woman's complaint was that he did not make love to her enough. The man had Mars in Scorpio in the Eighth House, square Pluto in Leo in the Fifth House. He would frequently play the guitar for hours on end. In effect, he was transmitting his sexual/emotional energy to his guitar. He was making love to it. Because of the fixed nature of this aspect pattern, he was unable to change his focus. In the end, the women started having affairs to meet her own legitimate needs. This produced emotional shocks for the man. Rage, anger, and violence followed. He physically assaulted his wife and her lover. He was arrested and jailed, and the women trotted off with another man. This emotional shock produced the necessary confrontation that made him examine where he was coming from, and for what reasons. His compelling desire for recognition, fame, and adulation with respect to his musical career, transmitted through the Eighth House Mars in Scorpio, translated into his "love affair" with his guitar. The guitar symbolized the goal and means through which he could realize his Fifth House Plutonian desires.

Others with Mars/Pluto aspects will compulsively need sexual release through relationships. The need and desire is to unite one's energies with other sources of power so that a transformation of personal limitations can take place. Naturally, then, these individuals desire to be sexually united and consumed through or with others. The quality of sexual expression is intense. When partners are not available, however, the pressure of the emotional/sexual energy can also translate into compulsive, yet necessary, masturbatory activity. The sexual expression is intense in these individuals because of the subconscious desire to penetrate and transform the limitations of themselves and others. These individuals have a need to penetrate and be penetrated. Some will use sexuality as a form of control or manipulation in relationships when emotionally wounded. For others, the bottom line of their lives will revolve around their sexuality — they will be consumed and driven by it. Most will experience instinctive and hypnotic sexual attractions to others Some of these attractions may turn into obsessions. Some individuals will claim the right and freedom to pursue these attractions, others will attempt to control the attractions by not following them, or not acknowledging them.

Many Mars/Pluto individuals will need to reject and rebel against sexual taboos as dictated by societal or familial environments. Some will explore or desire to experience many secret or "forbidden" forms of sexual experience. The power of sexual energy, and the desire for self-discovery through this kind of experience, can hypnotically attract these types.

Some with this aspect have experienced very painful and difficult emotional/sexual experiences in prior lives. Some have experienced rape, or have been the rapist. Others have suffered physical and emotional violence, or have been the violent one. When these conditions are indicated in the birth chart, or through the actual experience of the client, then the cause for these conditions will always be found within the individual's own prior-life background, or through causes that have been put into motion in this life. Most commonly these individuals have misused their emotional/sexual expression and energy in the past. This could have occurred through using their sexuality to get what they wanted from someone, or through emotionally manipulating a situation. In some cases, they have struggled with conflicts over commitment to a partner versus the desire to be free to be involved with whomever they wanted. This collision of desire can also be the basis of physical and emotional violence when one partner has felt used, manipulated or abused.

Other Mars/Pluto individuals have drawn to themselves partners who have had this problem of conflicting desires. In this case, that type of partner is only reflecting the overt or covert conflict inherent in the individual. The point is that some of these individuals have difficult sexual/emotional karma that has built up over time. The effect of sexual and emotional confrontations is to induce the awareness and knowledge of why these experiences are occurring. The answer always is found in the individual's own desires and prior activity based on these desires.

A few individuals have created an evolutionary situation wherein they have learned to deny or suppress their sexuality and needs. The fear of vulnerability, of being possessed or controlled by another, is the cause. Such people will attempt to transmute the energy toward the fulfillment of a goal or ambition, or they may masturbate as the intense buildup of energy demands a release.

In general, Mars/Pluto individuals have the capacity for total self-transformation, or total self-destruction. Of course, these two archetypical dynamics can cyclically alternate throughout life. Negatively expressed, the Mars component can become intoxicated by its own egocentric and individual power as a transmitted reflection of the separating desire of the Soul (Pluto). In varying ways, the individual refuses to acknowledge that anything or anyone is more powerful than him or herself. In this reaction, the individual refuses to accept any limitation. The attitude is commonly "I will do whatever I damn well please, and I don't care what the consequences may be." Such an individual can be amazingly ruthless and willful in obtaining or actualizing whatever needs or desires he or she has. The individual will challenge anyone who attempts to control or exert power over him or her, or anyone who threatens the "right" to do whatever he or she pleases. Underneath it all, the individual deeply senses the errors being made. At some point, the evolutionary desire emanating from the Soul to transform and correct the situation will manifest itself as undermining-type activity. In other words, the person will subconsciously set up situations or circumstances that will backfire. This effect will enforce emotional shocks. The emotional shocks will induce an inner Soul-searching that can lead to a metamorphosis of this type of behavior and personal identity.

For those who align themselves on a path of personal transformation, the rapidity of personal growth is rarely fast enough for them. Compared to the growth of others it is amazingly quick, but not quick enough from the individual's own point of view. Impatience, anger and rage are not uncommon reactions as these individuals encounter one limitation, one obstacle, one old compulsion or habit pattern after another. On the one hand they will desire to purge themselves of these limitations, and on the other hand they can become angry at the obstacles themselves and resent having to deal with them. This apparent paradox reflects the coexisting desires in the Soul. In addition, these types can be impatient at the growth rates of other people, especially those close to them. These behavioral dynamics can be intensified or mellowed depending on the overall nature of the individual, and the specific aspect from Mars to Pluto with respect to the houses and signs they occupy.

The greatest challenge in Mars/Pluto aspects is to align the personal will with a sense of higher will or purpose. Any resistance to this ultimate evolutionary need is based on the separating desires originating from Pluto that are transmitted to the egocentric structure of Mars. This resistance is based on fears of losing control, of being possessed or consumed by forces greater than the individual, and the sense of powerlessness that these fears induce. Once this resistance is eliminated, the individual's actions and desires will be in conformity to a higher will. Once this attunement to a higher will is realized, the individual can progress, attain and actualize his or her own purpose, desires and ambitions more quickly and effortlessly than do others.

PLUTO IN ASPECT TO JUPITER

In general, Jupiter/Pluto aspects promote an emotionally intense need to develop the intuitive dynamic or "sixth sense" that is innate to all of us. These individuals have become aware of the larger social/universal forces that constitute the totality of creation and have desired to understand the underlying natural principles or laws that are the foundation of creation. By developing an understanding of these natural laws or principles these individuals have attempted to understand their own sense of personal identity of individuality, their own natural laws, as connected to these universal laws.

These individuals are not concerned about intellectual opinions, or collecting information, facts and data for their own sake. Rather, they have been and will be concerned about developing broad, generalized and abstract concepts through which to explain these facts or details. In order to develop or actualize this need and desire these individuals have been learning how to let themselves become absorbed within the natural universal and social forces that surround and permeate them. Through this process of absorption, a natural silence or stillness manifests itself internally that allows for a transformation of the limitations in the intellectual function described by Mercury. Mercury represents the ability, power, and need to intellectually order and empirically (logically) connect life or reality on the physical plane. Mercury is concerned about physical laws while Jupiter is concerned with the metaphysical or cosmological laws that are the basis for physical laws. By allowing for an absorption to occur, these individuals can attune themselves to the natural cosmological or metaphysical laws that are the basis of reality in a generalized sense.

In this way, the intuitive dynamic has been developed. These individuals will be naturally intuitive coming into this life. They will naturally think in conceptual or abstract terms. The power of absorption promotes the development of intuition. Intuition leads to the condition of just knowing what they know without knowing how they know it. This knowing is not a product of deductive or analytical thinking, but simply occurs through pondering or speculating upon a question, problem, or the nature of things. By asking, pondering, or speculating in this way a natural absorption and linkage to the larger cosmological forces allows for the answer to come of itself. The absorption promotes an alchemical fusion through which the individual's consciousness is expanded by the natural cosmic forces that regulate the metaphysical laws themselves.

In essence, these individuals know what they know in abstract and conceptual terms. Frustration can be experienced by many of these individuals as they attempt to find the words that adequately capture or reflect that which they know. The frustration occurs when they attempt to communicate their knowledge to others. It is not an uncommon experience for these individuals to talk over the heads of others. They rarely do this deliberately, although many are accused of doing so. Many others may not have the ability to interpret correctly what they are trying to communicate. Yet, the transmission of this intuitive kind of knowledge can promote an understanding in others of what is being said without knowing how they understand it. In other words, Jupiter/Pluto individuals can spark intuitive development in others.

For some, the frustration of not being able to find the adequate words to communicate what they know leads to alienation and withdrawal from philosophical and intellectual interactions with others. Some individuals will develop an attitude of philosophical and intellectual superiority when they repeatedly find that others do not, or cannot, comprehend what they are trying to say. Some will develop an attitude of inferiority for the same reasons. Superiority promotes the need to convert others to one's own point of view. Inferiority promotes the need to learn even more thoroughly what one is attempting to understand in order to communicate what they intuitively know to be true.

In almost all cases, these individuals will experience philosophical and intellectual confrontations with themselves and others. Confrontations must occur for the following reasons: 1. to learn that their understanding of the natural, metaphysical or cosmological laws is relative or limited in some way; and 2. to learn how to communicate in language that is understandable and intelligible to most people.

Relative to the emotional power of Pluto, these individuals can forcefully communicate their beliefs and knowledge in a way that is often spellbinding in effect. Because they have developed their intuitive skills to tap into the natural laws that govern life, they will relate every-thing to those "bottom line" principles, truths or laws. These principles can be quite fixed and rigid in a Plutonian way. Such bottom lines corre-late to, and are the basis of, these individuals' deep inner sense of emotional security. These relatively fixed bottom line principles and convictions create philosophical limitations in most of these individuals. It's not that these philosophical principles or beliefs are wrong; in most cases they are right. The limitation is that these individuals do not always realize that there are other ways of interpreting or applying those laws and principles. The limitation can also derive from the fact that other natural laws or principles exist of which the individual has not yet become aware. Herein lies the need for confrontations in order to enforce the ongoing evolutionary purpose of eliminating or transform-ing those limitations so that growth and intuitive expansion can continue.

Confrontations are necessary to teach these individuals how to communicate in a language that is understandable. Jupiter/Pluto aspects promote natural teaching skills. Jupiter correlates to the wisdom or knowledge that we acquire as a result of our experiences throughout our lifetimes. In India, the planet Jupiter is called Guru (teacher). Relative to evolutionary capacity and karmic conditions, those with Jupiter/Pluto aspects have wisdom to share with others in some way. Yet, until they learn how to communicate in a way that can be grasped and accepted by most, this wisdom will go unheard or unaccepted.

Jupiter also correlates to the growth principle in all of us. Thus, these individuals have a particular need and desire to expand beyond the limitations of an imposed religious or philosophical orientation defined by their culture, family or themselves. Because the Jupiter/Pluto aspect, from an evolutionary point of view, desires to know the whole truth, these individuals need to challenge the limitations of any doctrine that denies or restricts the individual's right to experience or un-derstand more, or doctrines that deny the relevancy or legitimacy of other philosophical, metaphysical or religious system. This challenge can take the form of covert rejection, or overt attacks delivered or stated in varying degrees of intensity.

On the other hand, many Jupiter/Pluto people will themselves reject or deny other philosophical points of view because of their own emotional security needs. When confronted they will defend and main-tain their most cherished principles because of the emotional power of their beliefs, which are the basis of their security and sense of personal identity. In other words, they will use their rigid and fixed beliefs not only to defend themselves against external challenges, but also to rationalize philosophically the legitimacy of their defense. Living and dying upon their beliefs and principles, they will put down any system or point of view other than their own. These people may be leaders or followers of religious or philosophical sects that espouse the attitude of "We are right, and they are wrong." This same attitude can manifest itself in individuals who do not belong to any formal organization or sect, individuals who stand alone within their philosophical smugness rejecting all who do not agree with them.

In essence, those with Jupiter/Pluto aspects will intuitively realize and identify natural laws that are the basis of the reality that we share on Earth. In so doing, they can develop a belief system based on these laws or principles, and can conduct their lives in conformity to them. Most such individuals will have strong convictions and principles (morals) that not only dictate their own actions, but also influence how they understand and interpret life and reality around them.

I think most of us would agree that there are many and varied ways to understand and interpret the phenomena of life. Atheism, Buddhism, Christianity, Vendanta, Taoism, existentialism, nihilism, and so forth are all ways of understanding or relating to life in a larger way. Whatever path Jupiter/Pluto individuals are born into or are drawn to is based on their desires and karmic and evolutionary needs or conditions. Again, desires determine what we think we need. What we think we need determines the choices and actions we make to fulfill those desires. What we desire also determines what we value at any particular time. Our values and desires determine how we will intuitively and philosophically orient ourselves to life, and conditions what our beliefs have been and will be. The challenge for individuals with the Pluto/Jupiter aspect is to realize that each person's beliefs, convictions, and principles reflect personal truths in relation to their needs, desires, and values just as the Pluto/Jupiter person's beliefs are based on this same principle. Each person is oriented to reality in ways conditioned by their karmic and evolutionary development. If an individual's desires, values, and needs dictate an atheistic philosophical orientation, for example, then his or her intuition will tap into the natural laws, principles and truths regarding the nature and consequences of atheism. Another person's desires, values, and karmic and evolutionary needs would lead to some other philosophical orientation. Who's right, and who's wrong? Both are right relative to their needs, values and desires. With this realization, Jupiter/Pluto individuals can eliminate any defensive need to put down any other philosophical or religious orientation that does not conform to their own. They can learn how to teach, not indoctrinate, and can learn how to accept teaching from others. In this way, the search for the whole truth will continue.

PLUTO IN ASPECT TO SATURN

In general, Pluto in aspect to Saturn will promote the need in these individuals to actualize and realize their own authority, and establish this authority within society. Pluto can correlate to the largest sociological role that any of us can play within the context of society. When aspected to Saturn the evolutionary desire and intent is to fulfill or assume that role. This aspect promotes deep self-contemplation or reflection that leads to an awareness or knowledge of the individual's inherent capacities, and, therefore, the nature of his or her career or sociological role through which personal authority can be established and projected. The individual's capacities and role are relative to the houses and signs that Pluto and Saturn are in, the aspects that they make to each other and other planets, and by the natural evolutionary and karmic capacity of individual with this aspect. Once Saturn/Pluto individuals understand their capabilities, capacities, and the sociological role that they can play, they must also learn the means of actualizing these potentials as defined by their culture or society. In so doing, they will then be able to express their own authority with respect to the capabilities that can be expressed through the career or sociological role that they play.

From an evolutionary point of view, these individuals have and will be learning lessons relative to external authority and power. The culture or society itself is one source of external authority and power. Another source is the parent who yields most authority and discipline within the family. Since Pluto aspected to Saturn also correlates to the individual's need to assert his or her own authority, the aspect between these two planets will suggest how they have reacted to the need to express their own power and authority within the context of external power and authority.

The stressful aspects tend to promote a situation in which these individuals have challenged, rejected, or failed to understand the means of establishing their role, power and authority within their culture. They also suggest a clash or conflict of wills with respect to one or both of the parents who attempted to discipline the individual, or who expected him or her to conform to their standards of conduct. Because Saturn correlates to the phenomena of time and space, stressful aspects also suggests a problem with the discipline necessary to acquire the socially acceptable credentials that would allow establishment of their role and authority. Saturn/Pluto aspects demand that time be spent learning or preparing for the specific roles. The time factor also allows for the development of a socialized consciousness that learns how the "system" works, and what it is based on. By developing a socialized consciousness, the individual will know how to integrate his or her own authority and role within the society. Because the stressful aspects suggest conflicts or problems with this process in the past, many of these individuals will come into this life with a deep sense of frustration and futility. The sense of frustration and futility is based upon previous life experiences of being denied or blocked from being able to actualize their ambitions, goals, desires and sociological roles to the fullest possible extent. Denial or blockage occurred because they did not adhere to the necessary methods or ways to actualize these needs as prescribed by society. Thus, these people have an ongoing lesson to learn the proper methods in order to actualize their capabilities and needs to the fullest possible extent.

Commonly, the deep-seated basis of this prior refusal is founded upon a Plutonian fear of being consumed or controlled by forces larger or more powerful than the individual. This fear reflects a conflict between personal authority and the larger authority as manifested in the culture or the parent in authority. From an ultimate point of view, this larger authority originates in the Source that created the individual. The fear, however, is projected onto society. From an evolutionary point of view, the primary lesson is to learn how to submit, conform or align the personal authority to the larger authority and desire of the Source.

Others with a stressful aspect between Saturn and Pluto have succeeded in establishing their power and authority within society through the strength or force of their wills. But the means and methods used to achieve their ambitions and goals are commonly manipulative, ruthless, abusive and underhanded. Such individuals have or will experience emotional blows when the foundations of their career or role are forcefully removed at some point. The degree of emotional shock experienced will be in direct proportion to the level of resistance to changing their ways, or to the degree they defend their social positions and authority.

In either case, these individuals are continuing to learn how to accept the responsibility for their own actions. In the first case, these types face continuing lessons relative to self-determination, which is linked with an awareness of their capabilities, and the appropriate and required ways to actualize those capacities within their social role or career. Thus the individual's feelings of frustration are directly based on their own actions when those actions failed to conform to what had to be done to realize their goals. The evolutionary antidote is to do what is required of them to realize their capabilities in a self-determined way. In the second case, self-determination has been highly developed in relation to realizing ambitions, goals and capabilities. Yet, the means used to actualize them have been negative. Thus, these types must learn how to accept the responsibility for their own actions when the necessary confrontations, blows or shocks of removal occur, and to determine to correct previous errors. With stressful aspects, the understanding of the limits of personal and social power can be difficult to achieve. Emotional shocks and challenges delivered by others in positions of power and authority will serve to make these limitations evident.

Individuals with nonstressful aspects between Pluto and Saturn suggest that an understanding of the appropriate methods and ways to actualize their capacities and roles within society have been learned relatively easily. The nonstressful aspects convey an awareness of the appropriate role, and also a natural understanding of the limits of that role. They also convey an understanding of the limits of personal power and authority. These individuals will commonly understand the amount of time necessary to acquire the credentials that lead to the fulfillment and realization of their sociological role, and will naturally prepare for this role by creating a solid foundation upon which the role or career is established. In addition, they will also take their duties and responsibilities seriously.

All aspects between Pluto and Saturn contain the potential for leadership ability. The level of leadership will be determined by the individual's karmic and evolutionary conditions. Individuals with nonstressful aspects will accept and understand the obligations and responsibilities that go along with the leadership position or role that they are fulfilling. These individuals will also be able to accept and acknowledge the leadership abilities of others. In addition, they will be able to recognize and encourage the development of the natural capabilities of others, and not feel threatened by the authority or positions of others.

Those with the stressful aspects may not take their responsibilities and duties seriously. Rather, they may be more concerned with the position itself, and the maintenance of that role. This focus can promote an ultimate downfall if their duties and responsibilities are neglected or abused. These individuals may recognize the capabilities of others, yet feel threatened by them. Accordingly, they can act to suppress or undermine others in order to maintain their own positions. Those who have been or are blocked from being able to actualize their capacities to the fullest extent must learn how to reflect honestly upon themselves in order to determine the reasons and dynamics that have created this situation. In this way, they can become responsible for themselves and make the changes required to remove the blocks.

Regardless of the approach or attitude assumed by Saturn/ Pluto individuals, they all have the intrinsic ability to reformulate, eliminate or regenerate outmoded or stagnating dynamics or structures that are preventing further growth. This ability can be projected outward to make changes in their work or actions, or it can be internally focused in order to structurally change the orientation and parameters of their consciousness.

Saturn correlates to organization and definition. Every inner dynamic contributes to the internal organization and definition of our personal identity. Every outer dynamic contributes to the organization and definition of a society, culture or career, and to the totality of all forces as defined in a particular way at any point in time: reality. With Pluto linked to Saturn, the cycles of catharsis and metamorphic structural change will necessarily occur when evolutionary growth and development are blocked because of a dynamic, or a combination of dynamics, that are preventing necessary growth. This change will occur both within an individual and to the external conditions of reality: individually and collectively. The stressful aspects promote cataclysmic change in varying degrees of intensity, and the non-stressful aspects promote progressive, steady, evolutionary change.

In summary, then, Pluto by house, sign and aspect represents the most significant sociological role an individual can play. Saturn by house, sign, and aspect will suggest the appropriate means, methods, and ways necessary to accomplish and actualize Pluto's sociological role. The aspect between the two will correlate to how it has been or will be done.

PLUTO IN ASPECT TO URANUS

In general, Pluto in aspect to Uranus has and will promote lessons pertaining to the absolute elimination of all emotional, intellectual, physical and spiritual attachments that prevent or block growth.

This aspect promotes a subconscious urge, desire, or need to transform radically the individual's inner and outer environments. Strictly speaking, Uranus correlates to the personal or individuated unconscious dynamic in all of us. This individuated unconscious contains three kinds of information: 1. all prior memories, in total detail, of everything that we have ever experienced in this and other lives. 2. it contains what we repress about ourselves (issues, components, dynamics, feelings and so forth) that we do not want to contend with or remember, and 3. it contains the blueprint of what we can become in our most individuated state.

Uranus correlates to what Carl Jung called the archetype of individuation from a psychological point of view. This archetype represents the urge in all of us to break free from the external and internal conditioning that provides us with a sense of personal identity. Conditioning comes from a variety of sources: society, family, friends, lovers and our own memories of what used to be. These memories condition the very definition of our values, beliefs, attitudes, self-image and so forth and thus contribute to our attachments because of the familiarity and security that they represent.

With Pluto in aspect to Uranus the evolutionary desire, need and intent emanating from the Soul has been to eliminate all conditioning attachments that bind the individual to the past. The Uranus/Pluto connection correlates to the desire to be free. The three kinds of information contained within the individual's personal unconscious is either in a perpetual state of release to the conscious mind, or it is in a state of cyclic/cataclysmic release to the conscious awareness of the individual. The nonstressful aspects promote perpetual release, the stressful aspects promote cyclic/cataclysmic release. The contents of the individual's personal unconscious must be released into his or her conscious awareness so that the required freedom from the past can be realized.

The intent, effect, and experience of this release promotes continual or cyclic thoughts, images, or projections connected to the past and future. The interaction of the past and future takes place within the context of the present moment and induces an awareness or knowledge about how the past has conditioned the reality of the present. This awareness and knowledge thus creates desires related to the future: of what could be, of what they must do to be free of all that binds them to the past as reflected in the immediacy of their current reality experienced in the moment.

Reactions to the Uranus/Pluto aspects vary according to the overall nature of the individual. Some will be so radicalized as to rebel totally against all traditions, conditions, or limitations that pertain to the past, individually or collectively. The urge and desire to create new conditions or traditions promotes individuals, or a whole subgroup of individuals, who symbolize radical departures from the past. Many individuals with Pluto conjunct Uranus in Virgo, for example, created the Punk phenomenon. In rejecting the past, these people have created their own lifestyle (Uranus) and the symbols (Pluto) that represent the fact that they are punkers. Bonding together (Uranus in aspect to Pluto), they have individually and collectively rebelled against all conditions and traditions that symbolize the past, and in so doing, will transform the whole group (society) in some way.

Others will fear (Pluto) the future (Uranus). Thoughts or impulses related to the future will be perceived as threatening or undermining to their security, which is linked to the past (Pluto). Such individuals will attempt to identify with or link themselves to ideas, values, and beliefs that are associated with the past. Relative to the future-oriented impulses of Uranus, they will attempt to apply the past to the future in a new way.

Still others will simply identify and link themselves to future-oriented change as society changes with the times. Again, using the Pluto conjunct Uranus in Virgo subgeneration as an example, some of these people will follow the new conventions and traditions as created and defined by the mainstream of the subgeneration. In this way they will feel different because the subgeneration will be perceived as the leading edge of mainstream social change.

Throughout history Uranus/Pluto aspects have quickened the evolutionary growth cycles for all nations and peoples. Individuals with this aspect will experience a quickening of their own evolutionary pace and growth. They will feel an immediacy and urgency to transform limitations and restrictions remaining from the past. The forces of the past are accelerated as they rush to meet the forces of the future head on. If you study history you will find that the intensity and immediacy of the past colliding with the needs of the future can be measured by the type of aspect formed by Pluto and Uranus.

As an example, the Great Depression occurred as Uranus in Aries squared Pluto in Cancer. Hitler and Roosevelt both came to power in 1933 as Uranus and Pluto exactly squared each other. The resulting social changes and the course of history speak for themselves. At the end of 1944 and the beginning of 1945, Uranus and Pluto were in exact waning sextile. The intense conflict and stress induced by World War II began to ease as the seeds to end the conflict began to take root. The original evolutionary and karmic stress produced at the square began to manifest as a fundamental restructuring and organizing of the power bases in the world: the USA and the USSR. With this restructuring, the allies of these two nations polarized into blocs based on mutual philosophical, political, and economic interests. The seed that led to this reality was sown at the original square. The crisis in consciousness implied in that waning square translated into the opposing political and social orders that were symbolized by Hitler and Roosevelt. The process leading to the mutation effect of the waning semi-square through conjunction translated into a new world order with respect to the USA and the USSR.

Another seed that was planted to accelerate this new world order was the atomic bomb. Pluto correlates to the atom, and the penetration of the atom. Uranus correlates to inventions. The waning square through the conjunction promote a progressive expansion in consciousness that induces the awareness of the universal or Source. In this context, the knowledge that leads to the unraveling of the cosmic laws that allowed the atom bomb to be developed correlates to man's negative ability to play god. Amazingly enough, the primary inventor of the atom bomb, Oppenheimer, thought that if he built such a terrible weapon it would forever end the phenomenon of war. He was, by the way, a devotee of the Hindu goddess Kali. This goddess symbolizes death and rebirth to many people. Even through the 1950's most people thought nuclear technology would bring positive benefits: "Atoms for Peace" was the slogan while Uranus was in a balsamic approach to Pluto. When the conjunction of these two planets occurred in the early 1960's, following the showdown of Kennedy and Khrushchev, the masses began to realize the reality of what nuclear weapons meant with respect to the future. The waxing conjunction of Uranus and Pluto ushered in a new age.

Just as we can measure these cycles collectively, we can also measure the intensity, immediacy and manner in which the individual will respond to their own Pluto/Uranus aspect. The house and sign of Uranus, plus the aspects that it forms to other planets, will describe the vehicles and experiences that will allow these individuals to break free from the attachments that bind them to the past. The past will be symbolized by the house and sign placements of Pluto, the South Node, the planetary ruler of the South Node, and the aspects that these dynamics make to other planetary functions. The aspect between Uranus and Pluto will show how this break will be made.

PLUTO IN ASPECT TO NEPTUNE

In general, the Neptune/Pluto aspect will promote the need, desire and intent to realize the nature of illusions, delusions, dreams and ideals. In the deepest possible sense, this aspect promotes the need to dissolve any barrier or boundary that is preventing a direct alignment or relationship between the individual and the Creator. Yet, most individuals unconsciously fear letting go of the sense of egocentric individuality. To draw upon an analogy used earlier, the center of consciousness for most people is the wave upon the sea. Very few have their consciousness centered in the sea itself. Relative to the fear of dissolving into the cosmic sea, most people with Neptune/

Pluto aspects pursue one dream after another. Each dream being "the something" or "the meaning" that they have been seeking for so long. Yet, the evolutionary intent is to dissolve all barriers preventing the active and conscious aligning and centering of consciousness within the cosmic sea: to allow the Source or Creator to express Its will through the focus of one's specific individuality.

This aspect promotes the sense of an ultimate "something" or meaning to existence and reality: individually and collectively. Failing to identify what this ultimate meaning actually is, these people follow the dreams as if the dreams themselves are that ultimate something. It is precisely through this process of chasing dreams that the realization of the nature of personal delusions and illusions can occur. At some point, the dream is realized for what it is: a dream. Some will succeed in actualizing all of their dreams. Many will actualize some dreams and not others. A few will not be able to actualize any of their Neptunian dreams whatsoever. The dream, when actualized, may be meaningful and interesting for a time, but at some point it will lose meaning. Some people will quickly find out that the dream was not what they thought it was going to be.

In an archetypal sense, Neptune/Pluto aspects attempt to teach these individuals that they have the power (Pluto) to create and destroy reality at will (Pluto) if they believe in (Neptune) or desire something enough (Pluto and Neptune). The current popular interest in the power of visualization and affirmation techniques attests to this fact as reflected in the generation with Pluto sextile Neptune. There is even a current form of Buddhism with a specific focus and teaching revolving around this dynamic. Yet, most who follow this form of Buddhism do not actively cultivate a direct, conscious relationship with the Source through which this power is given to human beings. Most follow this path to actualize their deepest desires and dreams of a separating nature. One by one the dreams will turn into reality, the reality of ultimate emptiness and meaninglessness. This process can take a long time of course because dreams are limited only by one's imagination. Yet at some point the dreams will be exhausted. When this occurs, these people will face into the wind and, with faith and nothing to lose, desire to merge themselves with the Source that created them. As this process begins, they can become divinely inspired. Accordingly, the deep inner inspirations and dreams will be reflections of the Higher Will (the Source) that is guiding them in their own unique and individual way. Whatever these individuals create or become can have lasting value and meaning as a result.

From a spiritual point of view, this aspect produces a situation in which the Soul of each human being can be inspired by the

Ultimate Being. Collectively, all Souls are tuned into the Universal Inspiration via the signs and houses occupied by Pluto and Neptune. This factor produces collective evolutionary change in relation to the areas, experiences and dynamics correlating to the signs and houses that these two planets are in. Each individual will respond in his or her own unique way, yet be influenced by the collective generational vibration. In other words, each individual can tune into and dissolve the barriers preventing this spiritualizing effect with respect to the Pluto and Neptune placements in his or her birthchart. Each member of a generation does the same thing in his or her own way. Collectively, then, this general vibration will permeate the individual's social environment. The collective unconscious (Neptune) of the generation is attuned and receptive to the general evolutionary need for the planet.

The key point is that there is a natural evolutionary progression and spiral for the planet and the human species. This evolutionary progression is measured by the transits of Pluto, Neptune and Uranus. The signs that these planets are in will specifically correlate to where the evolutionary change is focused. All of us are tuned into this process in our own way, while affected completely by the social, physical, environmental and spiritual changes produced by the evolutionary impulse itself. The evolutionary impulse is Pluto. The collective unconscious is Neptune. The collective mind is Uranus. The collective need (Neptune) that reflects the evolutionary impulse (Pluto) translates into new thoughts or ideas that are reflected within each individual (Uranus). Collective groups of people are tuned into these evolutionary thoughts or ideas and for better or worse, collective change results.

A study of history provides dramatic evidence that reflects this evolutionary process, and how it fundamentally alters the planet and the human species. As an example, the industrial revolution evolved as Pluto was in Aries, and Neptune and Uranus transited Capricorn and then Aquarius. Today, with Pluto in Scorpio, Uranus in Sagittarius, and Neptune in Capricorn, the focus upon genetic engineering, as another example, will fundamentally change our world. Neptune in Capricorn presents the symbol of man playing God in this manner. The moral, ethical and philosophical questions (Uranus in Sagittarius) raised by genetic engineering can hopefully guide this development in the right direction. But who knows? Pluto/Neptune aspects promote dream inspirations that can be delusive and illusionary, or dream inspirations that can be in direct alignment with a higher will that is attempting to guide and shape the lives of us all, individually and collectively. Relative to the coexisting desires inherent in Pluto, the choices made by each of us, and the collective choices that are made by all of us, will determine whether the evolutionary impulse manifests itself as illusion or as true inspiration reflecting a higher will. Seventy-five percent of the people living today have Pluto in sextile aspect to Neptune. The realization that we are cocreators of our reality, relative to the choices that we make, is more critical than ever before in human history. The evolutionary and karmic consequences of those choices will be felt for a long time to come.

CHAPTER FOUR

PLUTO TRANSITS, PROGRESSIONS, AND SOLAR RETURNS

In this chapter we will be discussing the natural timing techniques of transits, progressions, and solar returns. This discussion will focus upon the natural evolutionary and karmic processes that coincide with Pluto and the nodal axis.

The natal chart, of course, represents the total potential of what the individual has been and can become. Transits, progressions, and solar returns correlate to the developmental timing of our life events. We will experience or manifest various dynamics at certain times because we are ready, in a karmic or evolutionary sense, to deal with these issues or dynamics. Clearly what we are working with is an evolutionary process experienced in two essential ways: 1. cataclysmic evolution and 2. slow, yet steady evolution. With respect to the evolutionary process, Pluto, more than any other planet, seems to prepare the individual for the necessary changes to come. Herein lies a key: choices can be made that will determine the intensity and quality of the necessary karmic and evolutionary events themselves.

Transits and Orbs

It has been my experience that the effective range of orb when dealing with Plutonian transits is five degrees. This orb also applies to transits of house cusps, planets in transit to Pluto, or transits to primary angles. Prior to the actual exact aspect to a planet, or to the exact crossing of a house cusp or angle, one is given the opportunity to prepare for the issues that will be occurring. These opportunities are associated with premonitions or forebodings about "things to come". Externally, signs or events of what is to come can manifest themselves on a very tentative basis. These signs or events will commonly be very irregular and disjointed, and have little apparent bearing on one's immediate reality. Of course, many individuals dismiss these signs, pay no attention to them, or are simply confused as to their significance and meaning

For those who have been traveling the path of self-realization, the capacity to understand or interpret these events and signs can help them prepare for what is to come. At minimum, these events and signs will catch their attention and they will attempt to understand what they portend for the future. Thus, some kind of preparation will occur.

This five-degree orb, prior to the exact aspect or house crossing, and after the exact aspect or house crossing has occurred, is similar to the process associated with volcanic activity. Internal forces build up to an eruption. After the eruption, the landscape is fundamentally altered. So too with the Pluto transit. The building process leading to and following the personal "eruption" is measured by the five-degree orb of Pluto's transit to a planet, house cusp, or a planet in aspect to Pluto.

The actual period of the eruptive event can be very slow or relatively quick depending on the speed of the planet aspecting Pluto. As an example, the transit of Mars to natal Pluto would involve about a three-week process. The events associated with planets transiting to natal Pluto will be generally less intense than the events associated with Pluto's transit to a natal planet, house cusp, or primary angle. Pluto itself moves very slowly, of course. Eighteen months is the shortest period for a transformative or evolutionary process, from beginning to end, when Pluto applies to another planet or crosses a house cusp. Or the process can take as long as five years depending on the resistance offered by the individual.

An example of this process is found with Richard Nixon, who has a Tenth House Pluto. His South Node is in Libra in the First House, and its planetary ruler, Venus, is in the Sixth House in Pisces. The North Node is in Aries in the Seventh House, and its planetary ruler, Mars, is in Sagittarius in the Fourth House in opposition to Pluto. The spectacle of Watergate occurred when Pluto was transiting Nixon's South Node via a conjunction, and inconjuncting the sixth House Venus. In essence, Watergate unfolded over an eighteen-month period from the initial break-in to Nixon's resignation. Yet, the inner process that Nixon himself went through in terms of understanding and resolving this event extended far beyond this eighteen-month period.

This example illustrates several key points. Watergate occurred through the cumulative consequences of prior actions implemented by Nixon. We all know the story, the deception, the abuse of power, and so forth. The event served as a catalyst that promoted Nixon's necessary karmic and evolutionary lessons. I think most would agree that Nixon did not consciously desire to resign as President. Yet, the very nature that he was born with promoted a defense of his social position: the presidency (Tenth House Pluto). Every action and activity that went into the defense of his position only intensified the eventual consequences leading to his resignation. Many choices and paths of action were open to Nixon that could have led to or created other outcomes. Yet his own actions, which were based on desires rooted in previous life patterns and orientations, created and lead to his fate: resignation and public humiliation. The emotional shock and cataclys-mic change were necessary in order to promote the awareness and knowledge of the deep inner patterns, dynamics and orientations that dictated Nixon's response to Watergate. As the Watergate episode unfolded, he had many preliminary signs and events that would have allowed for different choices to be made. Yet his old prior-life orien-tations dictated his responses to these signs and events.

In hindsight, as the process continued beyond the public event, Nixon seems to have made some very honest assessments of his behavior and the reasons for that behavior. It took his removal from public office to induce these assessments. His whole self-image had to be reformulated because it was no longer based on social position and power. His entire sense of personal identity and emotional security was essentially metamorphosed. He was forced to look at himself as never before. He finally came to accept and acknowledge the responsibility for his own actions. In my view, this step is extremely positive, for those realizations will fundamentally change or alter his approach not only to his own life, but also to social position and careers in this and other lives to come.

The Pluto transit functions the very same way for all of us. It brings into the light of day deep subconscious or unconscious emotional, intellectual, physical or spiritual patterns or dynamics that have been dictating and controlling our behavior and approach to life. Even though this process can be very stark and painful for some, the ultimate benefit is obvious because of the growth that will occur once these patterns and dynamics are changed or eliminated.

In this way, evolution occurs. Just as water will seek out the lowest point to the sea overcoming, undermining or wearing down obstacles in its way, so too will the natural evolutionary process over-come, undermine or wear down all obstacles that are preventing our natural and ongoing evolutionary growth.

Transits symbolize or correlate to what appear to be external cir-cumstances and events. These events have the effect of inducing inner change in all of us. Just as events and circumstances can range from the common and ordinary to ultra-intense, so too do our

inner changes range from "just cruising along" to total metamorphosis.

Too much emphasis in astrological circles and text books has
been placed on the transiting event itself, and not enough on the
cumulative process that has led to the necessity of the event. Growth
and knowledge occur through understanding all the issues and inner
dynamics that have led to the event. The point here is that an event
symbolizes the inner dynamics within all of us that have necessitated
the event itself.

If we can encourage ourselves and our clients to look at events
in this way, we can gain self-knowledge and take responsibility for
our experiences. Nowadays I think most of us would agree that nothing happens by chance or coincidence. However, there are many
people who feel that somehow things are just happening to them, as
if some big hand in the sky is making it so. Some will feel victimized
by events. Commonly, these are the people who do not have an
active spiritual life or much interest in programs or systems encouraging self-knowledge. Yet, even with such individuals we can help point
out the inner processes and dynamics that have created or led to
whatever events they have been experiencing. Clearly, if a person has
some degree of awareness as to what issues must be confronted,
and for what reasons, their attitude toward the necessary events will
help them to handle those events in a constructive way. Awareness
can help eliminate or minimize the common and characteristic
resistance to Plutonian changes, and even difficult experiences can
be consciously worked with to promote the necessary transfor-
mations in a positive way.

Pluto's aspects to natal planets promote the active evolutionary
process throughout a lifetime. The ways that this evolutionary process works are outlined in Chapter One under the subsection of The
Four Ways In Which Pluto Effects Evolution. Transits to or from
Pluto correlate to a specific time period through which the behavioral
orientation and manifestation linked to the planets will come under
the evolutionary process. Also, the houses and signs of the transiting
planet will contribute and be involved with this evolutionary process.
The transit of Pluto to other planets and houses, or planets in transit
to Pluto, correlate to the evolutionary and karmic needs and desires
of the Soul at any point in time. These changing conditions and
experiences are necessary in order to actualize, over time, the evolutionary and karmic intent for this life described by the natal position
of Pluto and the nodal axis. The transits that correlate to the necessary
experiences through time in order to promote the ongoing development of the inherent evolutionary and karmic needs will be experienced

relative to the four ways in which Pluto instigates evolution. It may be helpful for you to reread this section in Chapter One.

Plutonian transits will always fundamentally transform us in some way. Old and growth-inhibiting emotional, intellectual, physical or spiritual patterns, orientations, and dynamics will be eliminated or altered. New patterns, new dynamics and new orientations will replace the old. The transitional process from old to new can be very difficult for many people. The emotional security associated with the old and familiar patterns breeds resistance to change, thus the potential difficulty when change is required. The degree of resistance to the necessary evolutionary metamorphosis will determine the intensity or difficulty of the transition associated with the events themselves.

This natural evolutionary process is often associated with karma. In other words, the results or consequences of many previous actions and desires will come to a head at the time of Plutonian transits. The results or consequences can be negative or positive, or a mixture of both simultaneously. You cannot simply look at a birthchart by itself and make absolute statements regarding karmic results or consequences: i.e. Pluto transiting square Venus always means negative karma associated with relationships at that time. It may mean this, it may not. The experience of this transit would depend on the individual's previous actions and desires associated with previous relationships. Commonly, however, transits to or from Pluto will correlate to situations associated with the past relative to the planets and houses implicated.

It is not uncommon for an individual to be surprised by this situation. Many people will feel that they have already dealt with or eliminated those past conditions, associations, or dynamics. However, when this kind of situation occurs it will symbolize two things: 1. that the individual had not completely transformed or eliminated those conditions, associations or dynamics from his or her life, and 2. that the reason these conditions or experiences manifest themselves again is to promote a final purging or metamorphosis. It should be clear that the actual issues and events triggered by Plutonian transits are intimately linked to the context of each person's reality: to their natural evolutionary condition and intentions for incarnating and to their karmic necessities. The theme of a specific transit such as Pluto/Venus will be the same in all cases, yet the issues and events relative to the theme will be different. The person's approach and attitude will be different. The subtleties and nuances of the karmic and evolutionary conditions will be different in all cases, although the planetary theme is the same. Thus, the descriptions of Pluto's transit through the houses will illustrate the themes from an archetypal point of view. Each individual will react or respond in his or her own way.

How these themes are manifested in each person's life, again, is linked to the four primary ways that Pluto instigates evolution. In ninty percent of all cases, the Pluto transit tends to symbolize an ending, a culmination of a preexisting evolutionary and karmic line of development. However, out of the elimination or transformation of the old patterns, new seeds will be planted that will progressively bloom into new flowers of different forms and colors.

In about ten percent of the cases, the Plutonian transit will promote an awakening to new capacities and possibilities that were heretofore dormant or latent. The person will not experience an intense problem or difficulty requiring a transformation or elimination of some old pattern or dynamic. Rather, the individual becomes aware of greater possibilities and applications of who they are. The circumstances or events that surround the individual's life serve as the triggers or causes that ignite this awakening. An example of this process is former President Jimmy Carter. At the time that he inwardly decided that he wanted to become President, Pluto was transiting conjunct to his Twelfth House Libra Sun. The events and circumstances of that time triggered his internal response.

Progressions

If transits symbolize circumstances and events in our lives that reflect an inner process that leads to or creates the necessity of the event, what do progressions symbolize? Many astrological textbooks state that progressions symbolize inner or internal circumstances or events that lead to the necessity of adjusting, changing or eliminating certain external conditions in our lives.

In my view, this way of looking at progressions is limited because it fosters the notion that our external or outer life is distinct from our inner life. The fact is, one's inner nature or being creates or is reflected in outer life circumstances and conditions. Outer life conditions and circumstances reflect the inner person.

It is not uncommon in astrological practice to observe that progressions trigger external events. Nor is it uncommon to observe that many people report that they are inwardly feeling different in respect to some external condition in their lives when progressions are activating the birthchart. Both transits and progressions reflect the same principle that has been the basis for this book: the totality of our inner desires and actions, past and present, has lead to and created the external conditions of our lives. The outer conditions reflect the inner person from an evolutionary and karmic point of view.

None of us exists in a vacuum. Our consciousness, identity, self-image, values, beliefs and so forth are linked and conditioned by the outer circumstances of our lives. The culture, society, parents, friends and so on that surround us reflect our inner evolutionary and karmic needs and desires. The inner and outer are mutual reflections of each other.

There is no big hand in the sky making things happen to us. We make them happen ourselves. Self-knowledge reveals the reasons for that which is happening inwardly and outwardly. The reasons will always be based on karmic and evolutionary necessity. Considering the four natural evolutionary conditions or states, it should be clear that a large percentage of people are not aware that the inner and outer life are reflections of each other. Thus, many report or experience progressions as events. Those who have been traveling upon the path of self-knowledge and inner development will understand, experience and report how their inner changing dynamics are creating, or will create, external changes in the circumstances in their lives.

The experiential difference between progressions and transits, then, is based upon the evolutionary and developmental reality of each individual. In general, progressions will symbolize a situation in which a person will inwardly feel different about some already existing external condition in a way that may lead to changing that external condition, or the person may change some internal dynamic so they approach the external condition in a new way.

In general, transits will symbolize external conditions, circumstances, or events that promote internal changes, adjustments, elimination, or a metamorphosis of an inner dynamic. This process in turn leads to the change, elimination or adjustment of the external circumstances or conditions in the person's life. Again, in the last analysis, these two dynamics are one and the same.

Questions always arise regarding the orbs for progressions. I have found through my practice and experience that the effective time span that a progressed aspect impacts an individual's life is measured by a three-degree orb; one degree prior to exact aspect, the exact aspect, and one degree after the exact aspect. This process leads to change, causes the change, and then allows for reflection upon what the change has meant. The amount of time one will experience the evolutionary growth process symbolized by progression is dependent on the speed of each planet. As an example, Mars moving into a progressed aspect with Pluto would basically correlate to a six-year process because the average daily motion of Mars is thirty-five minutes. Thus two years prior to the exact aspect, two years during the exact aspect, and two years following the exact aspect. This example is based on the secondary method of progression (day for a year).

Planets progressed to Pluto will be metamorphosed in the four ways that have been mentioned. Just as a Plutonian transit, or a planet transiting with respect to natal Pluto, correlates in most cases to a karmic or evolutionary culmination, so too do progressed planets to natal Pluto. Just as in a few cases a Plutonian transit can awaken a person to higher capacities or possibilities that had been dormant or latent to that point, so too do planets progressed to Pluto promote the same phenomena.

Solar Returns

In solar returns the house position of Pluto, and the aspects to it from other planets, will correlate to the fundamental issues and experiences that an individual is dealing with in that year to help foster the evolutionary and karmic issues symbolized in the natal chart. The nodal axis in the solar return chart will show the modes of operation that will serve as vehicles to promote the evolutionary and karmic necessities and desires of the house position of the solar return Pluto. The planetary rulers of the nodes will facilitate this process. The house position of the solar return Pluto and the South Node with its planetary ruler will correlate to past experiences that will be experienced during that year. Some of these past experiences may have an element of karmic retribution, some will have an element of karmic fruition, and some will be based on distant prior-life causes. Whatever these experiences and issues are with respect to the solar return, they will be directly linked to the natal evolutionary and karmic conditions of the individual, and the evolutionary intent for this life. In the solar return chart the polarity point of Pluto, and the North Node with its planetary ruler, will serve as the path or vehicle to evolve, understand, resolve and transmute those issues and experiences. The lessons learned through these polarity points will have a direct linkage to the evolutionary lessons and intentions for this life. The aspects to Pluto and the nodal axis must also be considered in this analysis.

The Transit Of The Nodal Axis

The transit of the nodal axis with respect to the natal chart is also extremely important to the ongoing evolutionary and karmic development of the individual. Very few astrologers have given this transit much attention. This transit has a natural eighteen-year cycle to return to its natal position. This cycle, within other cycles, symbolizes in its own way that a chapter of an individual's life is ending and another is beginning relative to the nodal axis' return to its natal position.

Just as the natal position of the nodal axis by house and sign correlates to the modes of operation that have been used to actualize the prior evolutionary desires and intentions of the past (South Node), and the desires and intentions in this life (North Node), the house and sign positions of the transiting nodal axis at any point in time will describe contributing inner and outer conditions that symbolize, or are used to promote, the transition between the past and the future.

The transiting South Node, by house and sign, will operate as a channel through which inner and outer conditions or experiences in our life force us to deal with old attitudes, approaches, and modes of operation that are encountered in that area symbolized by the house and sign of the South Node.

Through counterpoint opposition, the North Node will operate as a channel through which inner and outer conditions in life will encourage us to develop new approaches, attitudes, and modes of operation not only pertaining to the house and sign of the North Node, but to also serve as the necessary new experiences that can help change the old approaches symbolized by the South Node.

The interaction of the past (South Node) and the immediate present (North Node) is continual in all of our lives. This interaction determines our future relative to the choices that we make based on this interaction. Just as transits to or from Pluto reflect the areas and behaviors that must be changed to foster the natal evolutionary intentions for this life, so too do the transiting North and South Nodes describe essential conditions that help promote the development of the natal modes of operation.

One last thought before we describe Pluto's transiting, progressed, and solar return meanings through the houses and to the planets: the reactions, responses, choices, and desires that we manifest throughout our current lives correlate to "karma in the making" — future karma. This karma is made in each moment of our lives in minor and major ways.

PLUTO THROUGH THE HOUSES

The description of Pluto through the houses will apply to both transits and progressions. The "inner" and "outer" conditions will be interwoven in these descriptions. When adopting these descriptions for the location of Pluto in solar return charts, apply them for the year that the solar return is in effect. Also keep in mind the natural ruler of each house, i.e. Mars naturally rules the First House, Venus the Second House and so forth. Thus, Pluto through the First House would correlate to Pluto transit Mars, Mars transit to Pluto, or Mars progressed to Pluto. If you use the solar arc progression technique, then it could also mean progressed Pluto to Mars if that progression was indicated at a certain juncture in an individual's life.

We will not be listing or describing aspect variations. Rather, these descriptions will simply involve the primary themes and archetypes. Thus, it will be necessary for you to apply the different and specific ways that an individual will experience the themes and archetypes relative to specific aspects. It may be useful for you to reread the descriptions of the specific aspects in Chapter Three.

PLUTO THROUGH THE FIRST HOUSE

The archetype or theme of this process is to initiate new actions and desires that reflect the fact that a new cycle in evolutionary development is beginning. This period of time normally creates an identity crisis as the individual feels deep inner urges to break free from the past that binds him or her, or is perceived to prevent growth. The past, of course, represents familiarity, and thus, security.

This new evolutionary impulse is now operating on a very instinctive basis. Deep urges to break free impact upon the individual's consciousness. These urges are not well formed thoughts. Accordingly, the individual has no real way of knowing where these urges will lead. Some will experience them in a very fearful way because they seem to threaten the existing reality and foundations of their life. No amount of rational analysis will provide the answers. In fact, the analysis itself can interfere with the process.

The key to handling these impulses positively is to simply follow the instincts themselves: if it feels right at a gut level, without one's knowing why, then to follow the urge to see where it leads. The understanding of why will occur after the fact. Conversely, if the instinct or urge feels wrong at a gut level, the key is to not follow them. The why will become known after the fact. Following the inner instincts and impulses in this way will lead the individual into the discovery of new paths and experiences that allow for the actualization of new dynamics and dimensions of the personality.

A common reaction to this process is to feel angry, anxious, restless and impatient. The individual may inappropriately project these feelings upon others or upon situations perceived as restricting growth. Or the individual can be angry at him or herself for the self-created conditions that now lead to a sense of containment and confinement. Anger can manifest itself for no apparent reason, anger that has been suppressed in this or other lives. It is not uncommon during this evolutionary process to draw circumstances into one's life that arouse anger, or to draw others who suddenly exhibit angry reactions toward the individual for no apparent reason. When these conditions occur the most common cause is unresolved or residual karma with those involved in such interactions.

From an evolutionary perspective, these circumstances or conditions create an opportunity wherein the individual can uncover and experience the basis of his or her anger. Accordingly, they offer the opportunity to purge this anger from the individual's psyche. In addition, a karmic opportunity is at hand to resolve old issues in which anger is being received from or projected upon other people. In worst-case scenarios, physical violence can erupt. The individual may be the victim of an attack, or be the attacker. If so, this kind of situation will be connected to some residual karmic issue rooted in prior actions.

It is not uncommon during this evolutionary process to experience problems in relationships, both intimate and casual ones. Problems will always occur through relationships that prevent growth for the individual. It is not uncommon, when the individual fears and resists the necessary process of initiating new actions and desires leading to independent self-discovery, for a partner, mate, close friend, or parent to induce confrontations that force the individual to break free and become more independent and self-assertive. In some cases, a partner may leave or may even die to induce this necessary growth process.

On the other hand, if the individual consciously desires to break free in some way, but perceives the partner, friend, or parent as holding him or her down, then the individual will initiate the confrontations leading to a potential break. If the relationships are not preventing growth, then it is up to the individual to implement the instincts and urges that they feel are right on a gut level. A few people will even have friends and partners who will encourage this process.

In addition, this process can promote the experience of being "hit on" by others who find the individual attractive or compelling. In certain cases, old lovers or friends may reemerge in the individual's life to prompt emotional, intellectual, physical or spiritual reactions that lead to new discoveries regarding his or her inner dynamics and needs. There are many reasons the individual can experience being "hit upon" by others who are expressing interest and attraction. For example, the individual may feel something is lacking within an existing relationship. The inner experience of lack creates an inner vibration that is projected into the external environment. Like throwing a fishing line into the water, the individual attracts others who will reflect his or her desires and needs that are not being fulfilled in the existing relationship. If the individual feels a gut response to follow such an attraction, then he or she should also be very honest and tell the current partner about it to avoid creating negative karma for all concerned. If the individual is unattached, then the key is to simply "go for it" if the attraction feels right at a gut level. The duration of this kind of liaison is not important. The quality of the experience is what counts. New discoveries and self-knowledge will occur. On the other hand, if the individual is already in a relationship that he or she feels is right for them, that supports the necessary growth, this kind of experience can simply serve as a test to determine their ability to remain committed, or it can serve to finish up old business in issues with the person(s) who manifest themselves in the individuals life. Again, there are many possible reasons for this kind of situation. An overall determination of the entire evolutionary karmic dynamics must be made in each case to understand why this is happening, and what to do about it.

Sexual confrontations and difficulties can also develop during this evolutionary process. The individual's sexual nature commonly intensifies during this period of time. Some will experience sexual limitations or problems associated with existing relationships that may lead to sexual attractions to others who may be perceived as filling unmet needs. For others, this process can promote hypnotic and instinctive sexual attractions, or the individual can draw others who are sexually attracted to them. These attractions can be very compelling. Some people will make new discoveries about the nature of their sexuality. In general, a metamorphosis of sexual identity will occur during this time. New experiences, directions, or expressions of the individual's sexual nature and identity allow for new dimensions of self-discovery. This process can be very positive if the individual follows the new impulses.

Physical problems may also develop during this evolutionary process. These problems can manifest themselves in the adrenal glands, muscles, blood (toxicity or unbalanced white and red blood cell counts), head, intestine, colon, liver, prostate, cervix, ovaries, womb, and the lumbar or naval chakra. Accidents causing physical problems are also possible.

Underlying these physical problems is the need to examine whatever is preventing growth, and to take the steps necessary for new growth. In so doing, health can be regained. The point is that the body has manifested a problem whose real origin lies in the individual's existing psychological structure and reality. These kinds of problems only occur when resistance manifests itself with respect to implementing the necessary life changes. Occasionally, one of the above-mentioned problems will develop in a person close to the individual. When this situation occurs the individual will learn these lessons through another by witnessing the effects of resistance to necessary change.

Whatever the specific events associated with Pluto moving through the First House, most people feel anxious, impatient and restless because they sense that something new must happen in their life. What is it? This question can produce inner fear of not finding whatever "it" is. Again, the key is to follow the instincts. Whatever "it" is will be discovered in that way. In all cases these new directions, instincts and impulses will involve the main karmic/evolutionary dynamic within the birthchart.

In general then, this evolutionary process symbolizes that the Soul wants to initiate and implement new directions, desires, and actions that will lead into new discoveries and dimensions within the individual. For some, this dynamic may require a complete break from the past. For others, it may only require breaking with certain aspects or dimensions that are linked with an old response, approach, or attitude from the past. In either case, the eruption will produce change in some or all areas of the individual's life.

PLUTO THROUGH THE SECOND HOUSE

The archetype or theme of this evolutionary process is one wherein the individual will experience a metamorphosis in how he or she is relating to him or herself and, consequently, to others. In addition, the process will promote a fundamental reevaluation of the individual's value systems.

The evolutionary impulse that this process induces is one of intense self-examination. Through inner confrontation the individual can eliminate and transform existing self-perceptions and the value associations upon which they are based. Because of this process, it is very common for the individual to experience an essential breakdown or loss of personal meaning. Life will not mean what it once did, and the inner sense of personal meaning will also change. Personal meaning is directly linked to personal values, which are directly linked to how one relates to oneself. This process demands that the individual confront the limitations in their value system in order to evolve and grow in new ways: to relate to him or herself in new ways, and in so doing, to relate to others in new ways.

The degree of intensity, conflict, resistance and confrontation relative to this process depends upon the individual's overall nature as reflected through his or her willingness or unwillingness to change. If the person's nature and value associations are relatively fixed, then the resistance can be quite strong with respect to this evolutionary impulse. On the other hand, if his or her nature is relatively open or mutable, then the resistance can be less severe. A simple clue to who's who can be found in the sign, house, and aspects of Venus in the natal chart. The planets ruling the Second and Seventh House cusps are additional indicators.

This process also promotes an evolutionary condition wherein the individual can become aware of new personal resources from within that can be used to establish either a new means of making a living, or to regenerate and reexpress the current means of doing so. In the first possibility, the individual finds it progressively harder to relate to his or her present way of making a living. This loss of meaning or relatedness induces the necessary confrontations so that the resources and values can emerge into the consciousness. In the second circumstance two conditions are implied: 1. the individual is still able to find meaning, value, and relatedness in his or her current livelihood, and 2. feels the inner need and desire to expand upon or reformulate the expression or application of that livelihood in order to reflect new dimensions of him or herself that are now being born into conscious awareness. In either situation, the experience of the inner and outer confinement or limitation promotes this development.

In some cases, this process will promote forced removal or depletion of existing resources, possessions, money, or the loss of a situation that is providing physical sustainment, such as a job, a marriage or partnership, and so forth. Conflicts over money or possessions can occur in some way. These circumstances occur because of the need to develop new values and ways of relating to oneself and others. These experiences also enforce evolutionary lessons of self-reliance, self-containment and self-sustainment. In cases of this kind, the emotional shock of loss induces or enforces self-examination in order to determine the reasons and dynamics that led to and created these conditions. For some the element of karmic retribution may occur. Such shocks induce inner isolation as, on a progressive basis, the individual cannot relate or give meaning to that which was lost. In this inner isolation the individual must learn how to respond to and develop new values and ways of relating to him or herself and others so that a new sense of personal meaning and value can develop.

In some way, the new personal resources can emerge through

which the individual effects a different means of personal sustain-ment on a physical and emotional level. From a karmic point of view, this situation is commonly associated with misuse, abuse, or manipulation of one's own or others' resources. In some cases, it is associated with denial or blockage of inner needs, capacities or talents that have not been acted upon. Thus, the forced removal allows no other alternative than the development and expression of new value associations that can lead to a new expression of personal relatedness and meaning. This in turn promotes new ways of relating to others, and allows for the actualization of latent poten-tials, desires, capacities and needs that create a new means of making a living.

In other cases, I have seen some individuals reap the rewards for sustained prior efforts in some area or aspect of their life. Some may even be rewarded for efforts in other lives. Rewards may come in the form of money, possessions, or recognition by others relative to the value of what the individual has been doing. In its own way, this condition also enforces new inner evaluation and self-examination as the individual must learn to give meaning to these new developments or conditions. Thus, the individual is learning how to relate to him or herself and others in new ways because of these new conditions.

Physical problems that may occur during this evolutionary process will commonly be related to the kidneys, lower back, the psychologi-cal function of hearing that can manifest itself as physical hearing problems associated with the structure of the ear, circulation dif-ficulties stemming from blockages in the veins or arteries, stones in the kidneys, urinary tract or gall bladder, cysts on the ovaries or womb stemming from extremes in the progesterone and estrogen levels, headaches stemming from a toxicity build-up in the kidneys and problems associated with the heart chakra.

In general then, this process symbolizes that the Soul desires to transform the existing value systems through which the individual gives personal meaning to his or her life. It is a time to eliminate all the old value associations, possessions, and ways of relating to one-self and others so that new patterns and associations can manifest. By allowing the well of the past to be emptied, the well of the future can only be refilled. Resistance to the process promotes not only psychological, karmic and evolutionary problems, but can also promote the above physiological problems which are all based on blockage. Remove the psychological and emotional blocks and the individual will metamorphose into a regenerated state of personal vitality and well-being.

PLUTO THROUGH THE THIRD HOUSE

The archetype or theme of this evolutionary process is one wherein the individual will confront the limitations of intellectual structures, organization and orientation. Thus, it will be necessary to examine the nature of personal opinions and the basis for those opinions.

As this process increases in intensity, the individual will progressively begin to experience an implosion of his or her intellectual constructions, logical connections, and mental orientations to him or herself and the environment. This implosion is commonly accompanied with intellectual disagreements and arguments with others. Subconsciously, the individual will draw circumstances and events that enforce or induce this evolutionary need.

These events and circumstances are normally associated with new people or experiences that confront or challenge the individual's opinions and the basis of those opinions. Events can also be associated with people in the individual's life who begin to intellectually reorganize their own lives and question the basis of their own opinions. Thus, where there was once intellectual agreement with certain people, there now is disagreement and problems.

The reason for these conditions is that the individual must progressively develop new points of view and expand upon his or her intellectual framework in order to experience and perceive life in new ways. In addition, this process must occur so that the individual can eliminate superfical or poorly thought-out ideas about the nature of things. In so doing, the individual can awaken to the underlying reasons of why he or she thinks about things in the way that they do. This in turn will promote the awareness of why the individual has intellectually organized reality in a particular way.

A deepening of the mind is now occurring. By examining the basis of opinions, mental orientation, and intellectual organization, the individual will progressively experience three states: 1. the need to eliminate outright certain opinions, mental attitudes, points of view, or a whole system of intellectual organization; 2. the need to reformulate or deepen certain ideas about the nature of reality that are still relevant and functional; and 3. the need to reach out for new experiences that allow for new ideas, information, and knowledge to come.

The common psychological symptoms that ignite this process are feelings of being intellectually cramped or stifled. Intellectual restlessness and boredom are also common. This process, of course, can threaten the individual's sense of emotional security, which is connected to his or her intellectual organization, ideas and opinions.

Thus, to change the ideas is to risk inner emotional insecurity. In some of these people this risk creates resistance to the necessary metamorphosis. The degree of resistance determines the degree of intellectual confrontation within and from outside. Again, clues to the degree of resistance will be found in the natal condition of Mercury, the signs of the Third and Sixth Houses, and the conditions of the planetary rulers of those houses.

The need during this process is to bring in new information from the "outer" environment, and to allow new informaton or ideas to surface from the inner depths. In this way, new perspectives will manifest themselves that allow the individual to examine and understand the basis for his or her existing intellectual beliefs and opinions. Because this process reflects a desire for deeper knowledge about the nature of life, inwardly and outwardly, it is not uncommon for the individual to feel repelled or irritated at the superficial explanations or opinions that others express. I have seen normally docile individuals erupt in a tirade of intellectual attack upon others whom they now perceive as shallow, mindless or simply fixed in their opinions. Of course, this outer process is a direct reflection of the inner eruption that the individual is experiencing.

It is a good idea, and not uncommon, for such people to seek out new knowledge systems and ideas with which to build a new foundation so that the necessary reorganization does not become overly dispersed in a variety of directions. It helps to become involved with or study one comprehensive system and discipline so that other interests can be related to the new foundation that is reflected in the one comprehensive system. The danger is to fly off in many directions simultaneously because of the deep inner thirst for new information and ideas. Scattering the energies will only deepen the intellectual crisis because now the individual will be lost and confused by a revolving door of ideas and perspectives that contend with or clash with one another.

On a preliminary basis, as this process begins, it can be advisable to sample or checkout a variety of new intellectual systems in order to discover the ones that most perfectly reflect what the individual needs for the necessary reorganization to occur. When the one system is found, a commitment must be made so that a foundation can be created upon which all other new ideas and information can be given consistent perspective.

Travel is another beneficial activity that can promote new thinking patterns. When one is traveling one is in motion. Being in motion allows for processing of thoughts. Traveling brings new experiences, contacts, and conversations with the people that are encountered along the way. The need for traveling is cyclic during this process. Too much traveling will promote a scattering or dispersal effect wherein the individual can lose all sense of a personal center. Not enough motion or traveling can promote an intensification of the natural intellectual implosion, which in turn can cause an intellectual and emotional loss of perspective and sense of the inner center. On the other hand, on a cyclic basis, a need for inner exploration arises wherein the individual discovers new thoughts, perspectives, and mental orientations by being still. The key is to be aware of these two natural cycles during this time frame: to put oneself in motion when necessary, and to rest or be still when necessary in order for a settling to occur.

Physical problems that can occur during this process are commonly associated with the nervous system in a general sense, neurological problems, rashes on the skin, lung ailments associated with the lining of the lung, throat problems, and problems associated with the cervical or throat chakra.

In general, this evolutionary process promotes the need to expand mental horizons and to eliminate outmoded thinking patterns and opinions. It is time to think things through in order to discover the basis of intellectual attitudes and assumptions. Personal knowledge and self-realization will come in this way.

PLUTO THROUGH THE FOURTH HOUSE

The archetype or theme of this evolutionary process is to induce inner and outer circumstances wherein the individual will find it necessary to examine the basis of what constitutes his or her emotional securities, dependencies, self-image, feelings and moods. In addition, this process will serve as an excellent time to examine the impact of the individual's early environmental situation as reflected through the parents, and for those who are parents, how they themselves have emotionally responded to their own children, family, and spouse.

As a timing device via transits, progressions, and solar returns, this process promotes the ending of a chapter or cycle in the individual's life, and the beginning of a new phase. It will now be necessary to change or eliminate all forms of emotional dependency and security that are linked to external situations. These dependencies and securities are in some way limiting further growth.

This evolutionary time frame and experience can be very difficult because many people will feel as if the very foundations of their lives are being threatened and removed. Such an experience must occur so that the individual is more or less left with only his or her self to look at, to examine, and to depend on. One's self-image is a reflection of the composite nature of one's total reality, inwardly and outwardly. When certain aspects or dynamics are removed or eliminated, the individual is forced not only to examine how the self-image has been conditioned and based on this fact, but also to now reformulate a new self-image that is based on necessary changes occurring within and outside the individual.

Emotional upheavals will surge forth from the inner depths, and will also be reflected in the outer environment through a variety of circumstances. These external causes or circumstances can range from problems associated with the personal living environment, family and parents to job or career, and other closely associated people. The effect of these upheavals can be quite unsettling for the individual.

The evolutionary intent and reason for these inner and outer conditions is to force examination of how one has put life together, and for what reasons. Within this intent, it is also necessary to examine the basis of emotional security and self-image relative to the nature of personal reality. Accordingly, it is now time to change old patterns of emotional response to oneself and others. In addition, it is now time to change or eliminate all sources of external dependencies relative to personal security at an emotional level. For some people, this process will be enforced through the loss of a job or career, a family member or someone close to them, emotional confrontations of an intense magnitude with family members, or even the loss of the individual's own life. For some, the emotional fear of death may become a preoccupation. Others may experience a close call or brush with death.

This process allows for an emotional cleansing or healing through the release of old emotional patterns and attitudes that have built up for many lifetimes. Some of these patterns are also related to emotional needs or responses that have been stifled, denied, or buried for a long time. As these emotional patterns surge forth into the conscious awareness of the individual, the moods and feelings that are a reaction to these emotional winds can be extremely intense and quite compelling. Each shifting and passing emotion will be capturing the total attention of the individual for its duration. These emotions, moods and feelings promote conscious awareness as to the nature of the individual's old unconscious and habitual ways of emotionally responding to life. Certain people will appear almost catatonic because of their absorption within these passing emotional states. Such absorption must occur because it serves as the vehicle through which the individual can totally examine the basis, causes and reasons for his or her emotional responses, needs, demands, security equations, self-image, and the impact of the early environment.

The conscious awareness and knowledge that this examination induces can lead the individual into the experience of emotional and personal stagnation. The sense of stagnation is necessary, however, for it promotes the opportunity to change. Family members or others close to the individual may also experience these same conditions in their own lives. Unresolved emotional issues associated with prior-life causes or conditions can now surface between the individual and those in close relationship. The behavior associated with these old unresolved emotional issues can be quite irrational and combustive. On the one hand this can be an excellent time to resolve whatever those issues are about, and on the other it can promote a termination of the relationship if the individuals find it too difficult to work through and resolve those issues. In either case, new emotional patterns, self-knowledge and necessary change will occur.

During this process the individual should be encouraged to reflect upon the past with conscious intent. In other words, to reflect upon all events, situations and circumstances in the life to date, and to examine the inner dynamics or causes that necessitated or created those conditions. In this way, the individual can become more aware of his or her inner nature and structure.

This process of active reflection should also be linked with active communication or processing. This cyclic or rhythmic need to release or communicate that which is being discovered from within is necessary because it will allow for the essential purging or cleansing of the emotional structure. In other words, a danger exists wherein the individual can emotionally implode and distort if he or she remains in a withdrawn or reflective state. This rhythm or cycle will not be predictable, but will fluctuate on a daily or even moment to moment basis. The individual should simply be encouraged to release, process, or communicate as necessary. This process requires at least one other person in the immediate environment whom the individual can trust, and who is mature enough to allow this process to occur without feeling threatened or blown away by it, to be present.

The issue of whom to trust can be very difficult because many of these individuals will not even be able to trust themselves in terms of what they are feeling or needing. Some will feel intensely vulnerable and super-sensitive during this process. These emotional needs at times can be very similar to the needs of a child, no matter how old the person may be. Since the intent of this process is to learn inner security and to minimize external dependencies, those in the individual's environment who are trying to help should gently promote this understanding.

Physical problems that can manifest during this process will be associated with the lymphatic system, stomach, mammary glands, eyes, mucous or water retention, toxicity buildup in the mucous, and constipation, as well as problems associated with the medulla chakra.

In general then, this evolutionary process promotes a total rebirth or metamorphosis of the inner emotional structure and self-image of the individual. By eliminating and changing all the old emotional patterns and minimizing external dependencies, a new person is born. A new chapter is beginning in the individual's life.

PLUTO THROUGH THE FIFTH HOUSE

The archetype or theme of the evolutionary process is to induce an inner and outer situation through which the individual progressively learns to take charge of his or her own reality, and to recreate his or her reality in order to reflect and establish the new creative impulses that are emerging from the depths of the Soul. These new creative impulses will be directly linked to the very purpose of the individual's life.

In addition, the process will necessitate that the individual examine the basis of why and how he or she has been giving to other people. Within this evolutionary need, it will also be necessary for the individual to examine the basis of the personal need to receive love and attention from others, and to be considered important and special. With these two necessary lessons combined, the individual can now experience and perceive how self-centeredness conditions the manifestation and application of each need.

In general this process will promote a regeneration of personal creativity. Deep inner emotional surges of creative inspiration, feelings and desires will impact upon the conscious awareness of the individual. The new thoughts, feelings and desires will spur the individual to take charge of his or her destiny and personal reality. The effect will be to recreate personal reality in order to accommodate, apply and express these new creative surges. The very purpose for the individual's existence will be reformulated as a result. This can be a time wherein latent or dormant capacities can now be developed.

During this process some individuals will necessarily become and/or appear more self-centered and self-consumed than ever before. For others it will require the breaking up or elimination of an overly self-centered focus so that the new impulses can be expressed. The key in either case is to allow oneself to move toward these new

impulses as they occur, even if they demand letting go of preconceived ideas or beliefs as to what the individual identified as their
purpose and direction in life prior to this time.

The inner evolutionary impulse will also be reflected in outer
conditions and circumstances. The nature of these external conditions will depend on the specific nature and context of the individual's
reality up to this time. Let's use the example of having a child to
show how all these principles could manifest. For some people these
circumstances and conditions will be associated with having a child
as a reflection of personal creativity, giving and receiving, and a
reformulated purpose in life. In these circumstances the issues of
giving and receiving will occur through the new baby. Self-centeredness
can occur through narrowing the individual's focus to the needs of
the baby and oneself. For some, having a baby will provide a vehicle
to eliminate or change excessive self-centeredness and narcissiastic
behavior and orientations. For others, pregnancy may occur through
a love affair or an extra-marital affair. Others will become pregnant
without having a conscious desire to do so. In all cases, there is a
karmic connection with the baby and a karmic implication for the
father as well as the mother. When Pluto begins its transit through
the Fifth House, or when it is transiting the Sun, progressed to the
Sun, Sun progressed to Pluto, or Pluto in the solar return Fifth
House, an increase in fertility will normally occur.

In other cases, the individual can experience emotional confrontations and difficulties with his or her own children. Just as the
individual needs to develop a new creative purpose, application and
identity, to take charge of his or her own destiny, so too will his or
her children begin to experience and manifest themselves in this
way. This dynamic may pose a challenge or test of wills between all
involved. When these conditions occur, it will almost always reflect a
situation wherein the individual had been attempting to mold the
identities of the children to fit his or her own image of who they
should be. As the children begin to assert their own wills and
demand for freedom to follow their own desires, they will directly
challenge the individual's own desires of what he or she wants the
children to do. This confrontation will promote the evolutionary need
to examine the basis of how the individual perceives and gives to the
children. Through confrontation and a clash of wills, objectivity can
occur wherein the individual reformulates his or her image of the
children to reflect their actual needs and inherent individuality. In so
doing, he or she can now give to them what they actually need. This
process or condition will also reflect itself through the creative redefinition of the individual because of the changes, adjustments and

reorientation that this circumstance or condition demands.

During this evolutionary process the individual can also become attracted to others, possibly leading to a love affair, especially when the person is feeling unloved, unfulfilled, or unacknowledged within the existing conditions of his or her life. Because of the deep inner urges to become more creative and to take charge of personal destiny that are welling up from the Soul, the individual can now be subconsciously attracted to those who help fulfill and promote these needs. A love affair can satisfy the need for recognition and the need to be considered special and important. Such attraction during this time will be intense and spellbinding. Those drawn by the individual will be very magnetic and powerful and others will perceive the individual in this light, as well. Some encounters will be brief though intense and others can be long lasting, even developing into a life partnership.

If the individual is already in a relationship, this situation can certainly threaten it. Some will feel completely torn and divided. They will be in love with both the new person and the existing partner. The basis of this situation, again, is that the individual was not feeling enough love, attention, or recognition from the existing relationship. The individual may have also experienced denial or repression of his or her new creative needs and purposes in the existing partnership.

Such a situation will subconsciously "set up" the opening to become involved with or attracted to another who will fulfill these needs. Either the existing partner must change his or her own emotional orientation to allow for and encourage the individual to actualize whatever is needed in order to express the new life purpose, or the relationship will most likely be terminated at this time. This potential event would symbolize the need to take charge of the personal destiny so that the individual could recreate it in such a way as to reflect the new creative impulses emerging from the depths of the Soul.

Physical problems that can occur during this evolutionary process would manifest in the heart, circulation and blood flow, decrease in constitutional strength or vitality, sharp fluctuations in energy levels because of stress, boils, abscesses, tumors, and problems associated with the ajna chakra — the "third eye."

In summary, this process promotes a reformulation of the individual's purpose in life. New dimensions of personal capacities and possibilities will surge forth from the depths of the Soul. By taking charge of his or her personal destiny, the individual will recreate the self in some way *if* he or she can follow the inner creative urges and outer environmental feedback as they manifest. This process may require elimination of preconceived ideas and beliefs about the life purpose and destiny. If the individual stubbornly resists going with the new order, then he or she will be blocked and stifled from further development and creative expression. Thus, rather than experiencing a creative renaissance, the individual will experience a decay and blockage of creative expression.

PLUTO THROUGH THE SIXTH HOUSE

The archetype or theme of this evolutionary process is to induce a period of time for personal improvement. The need for personal improvement will manifest as a deep inner self-analysis through which the individual mentally scans and surveys all the components and dynamics within him or herself that need to be adjusted, changed, or eliminated outright. Thus, the individual will experience an essential period of time in which self-criticism permeates the being.

It is a time of cleansing, purifying and healing of the Soul and physical body. Initially, however, conditions may get worse before they get better. A deep inner sense of impurity, impefection, inadequacy and negative feelings about oneself will be reflected in the outer environment. The person will now perceive outer conditions exactly the way he or she perceives inner conditions. This projection, of course, attracts the very same vibration back to the individual via the environment.

Experiencing inner and outer criticism, a sense of lack, inadequacy and imperfection will induce deep inner feelings of doubt. This doubt is a necessary experience because it will allow the individual to change and adjust the inner and outer environments to reflect new ideas, thoughts, feelings and needs now manifesting themselves from the depths of the being.

It is a common experience during this time to become consumed or involved in inner and outer crisis. The inner crisis derives from the necessity to let go of preconceived ideas, beliefs, attitudes and mental equations of all kinds. The need for change will impact on the emotional, physical and intellectual structure of the individual, and can undermine or threaten the individual's sense of stability and security.

Up until this time, the individual has held certain attitudes toward his or her work, physical body, and the mental organization of inner and outer reality. We are dealing with the Virgo side of Mercury in the Sixth House. Where the Gemini aspect of Mercury needs to collect a variety of information in order to make logical connections that empirically order the physical environment, the Virgo side needs to analyze and compartmentalize that information; to put it in boxes and classification systems. The focus is on facts and details; how one fact or detail connects to the next. By making such mental connections, a composite picture as to the nature of reality is created. It should be clear, then, that the particular facts or information that the individual selects or focuses upon will limit and condition how the person perceives the total picture of reality. The selection of facts and information taken in is influenced by the individual's prior evolutionary condition and development. The crisis will thus revolve around the need to reorient the mental and emotional focus to bring in new information and knowledge. The old mental equations and attitudes about what constituted reality are now breaking down; they just won't work anymore. Thus, a deep inner cleansing of body, mind and Soul occurs at this time due to the purging of the old mental attitudes that have become barriers to further growth.

The individual will experience a deep inner doubt based on the limitations and inadequacy of these preconceived ideas. The doubt will induce self-analysis, which will promote the mental scanning of what needs to be changed or eliminated. Self-analysis will promote the awareness that these preconceived ideas and mental equations are deficient in some way, and that new information, attitudes and approaches to inner and outer reality must be sought.

In effect, the past and future are now colliding in each moment of the individual's experience. This collision induces the crisis because of the implied resistance of Pluto. The resistance will manifest itself as mental stress that promotes anxiety, irritability and nervousness. Such mental stress will commonly manifest itself as ailments in the physical body.

Outwardly, the individual can now experience difficulties or confrontations with others who may be critical toward him or her. This criticism will be based upon what they perceive to be wrong with the individual. The work environment can also be difficult now. The individual may feel deep inner desires to change the level or nature of work, the approach to work, and may not feel appreciated for the quality of work done. The work itself may be perceived as limiting and stifling further growth. Employers or employees may promote confrontations or difficulties. Some individuals will lose their jobs during this time through karmic retribution of some kind, or to enforce the necessary changes.

The recommended strategy is to accept and reflect upon the nature of external criticism, and not to feel defensive, angry, or bitter because of it. The point is that the outer environment is reflecting the inner, and these messages are necessary clues and indicators as to what the individual must look at. Thus, this process will promote a positive and necessary cleansing of body, mind, and Soul.

If the individual feels the inner draw to change his or her work, it will be important to create the necessary foundations that will allow for a gentle or gradual transition between the past and future. The person might secure a new job prior to leaving an existing one, or might get new or additional training while maintaining the existing job.

If the individual is experiencing physical problems, then it may be advisable to change nutritional habits and the kinds of food eaten. Because this period is for cleansing, lots of fluids, raw vegetables, cultured yogurts, fruits and grains would be good choices. Refined or artificial forms of sugar are best eliminated. Hot baths, yoga and massage would be beneficial.

Just as the old mental attitudes are being purged from the Soul, so too are the old ailments and debris built up over many years. While the process is natural at this time, mental stress creates the trigger that ignites physical ailments.

Physical problems that can occur during this evolutionary pro-cess can manifest themselves through genetically based ailments that have been dormant or latent, or nagging problems that now become intensified. Problems related to the pancreas (insulin and enzyme production), intestines, colon, liver, spine (loss or deficiency of necessary fluid in the spinal cord), nervous system and hearing can also develop at this time, as well as problems associated with the throat chakra.

In summary, this process represents a time of personal improve-ment sparked by a self-examination of old mental attitudes and equations that need to be changed or eliminated. As a result, the individual experiences a cleansing on all levels. Cleansed and purified, the individual can emerge from this period into a state of intense personal vitality, with new approaches and attitudes toward life, and an excitement at continuing on with his or her life journey.

PLUTO THROUGH THE SEVENTH HOUSE

The archetype or theme during this evolutionary process reflects an essential need to transform, change or evolve attitudes, values, values, needs and approaches pertaining to close relationships or partnerships.

As Pluto approaches the Seventh House cusp, begins a transit to Venus, or Venus by progression aspects Pluto, a karmic time frame exists wherein the individual may reexperience others with whom there have been past life associations. This initial karmic time frame can also bring renewed connections with people from earlier relationships in this life. Occasionally the fleeting Venus transit to natal Pluto can also induce this situation for a few months. The specific kind of aspect will generally indicate the karmic basis of the contact. The stressful aspects will tend to promote the need to deal with unresolved issues between the two people. The nonstressful aspects will tend to promote a situation wherein the individual draws others who have positive and timely information to bring to him or her at this time.

In either case, these conditions reflect the fact that the individual desires and needs to move on; to change, grow and evolve into new ways of being in a relationship. The need to change the past ways of relating, of redefining one's essential needs, is reflected in these conditions. The need to reevaluate draws these types of people into the individual's life. Those that promote confrontations due to unresolved issues induce the necessary awareness of past conditions that need to be resolved and changed. Those that bring timely messages reflect the same thing. The individual will be open to those people who bring these messages or who reflect what he or she is desiring to become. Conversely, the individual can manifest resistance to dealing with those who symbolize the past and embody old unresolved issues. Because of the need to move on, all situations and people who represent the past will be perceived as being in the way of the new, the future. Thus, conflict promotes a resistance to dealing with the karmic factors implied in these contacts. Yet, they must be resolved because a new chapter or evolutionary cycle is unfolding in the individual's life. The past conditions reflected through these contacts must be resolved and fulfilled so that a new cycle can be born and manifested without the residue of the past contaminating the future.

From another point of view, the confrontations induced through such contacts will promote an experiential awareness of the new needs within the individual. The experiential confrontation thus allows for an alchemical metamorphosis to occur through which the new needs, attitudes and approaches to relationships can evolve into consciousness. Each case that reflects these conditions is unique. The specific nature or karma of what is being resolved between the individual and the other will be found through cross chart comparison (synastry).

In my view, the individual should be encouraged to fulfill and resolve the issues so that the karma with others can be finally eliminated once and for all. I recommend a psychological strategy wherein the individual resists the temptation to react instinctively in a confrontational way to those embodying the karmic past. The key is to develop an attitude of considered response, to view the situation in a dispassionate and reflective way. The individual can thus become aware of all the issues that have led to and created the situation, and can respond in appropriate and karmically necessary ways to resolve those issues.

In some cases this process can also bring into light old unresolved issues between the individual and an existing partner. These issues may have been repressed earlier in the relationship, or they may resurface as unresolved problems from other lifetimes. These problems or issues can produce confrontations between the individual and the partner. In these cases, the need is not only to deal with these issues as they occur, but also to allow the existing relationship to change, grow, evolve and redefine itself to reflect the new needs of each partner. The same strategy of considered response should be encouraged so that the necessary evolutionary metamorphosis can occur along positive lines.

In some cases this process will lead to the termination of existing partnerships if they cannot be redefined because of compulsive resistance to the new evolutionary demands. One individual may perceive the other as being in the way of future needs. This can be the individual him or herself, or the partner with whom he or she is involved. The other will perceive the individual who is demanding change as undermining and threatening existing emotional security. Whatever role or position the individual finds him or herself in, growth will occur even though these experiences can be very intense and difficult. Occasionally, this evolutionary process will correlate to the death of a spouse or someone close to the individual.

The nonstressful aspects will also stimulate growth. The main difference is that the individual, and those that he or she is dealing with in these necessary ways, will commonly understand the reasons for the experiences. With stressful aspects, the understanding occurs after the fact. The results, however, can be the same. In whatever way this evolutionary process is experienced, the need is the same: to eliminate or change past patterns of relating to others, and to allow the new essential needs to surface into the consciousness. New ways of relating to others, new personal and social values and new ways of relating to oneself will unfold at this time as a result.

It is not uncommon for new people to manifest in the individual's life at this time. These new people will generally be very powerful and intense, and will tend to have a hypnotic and compelling effect on the individual because they will symbolize his or her new needs. Others can now perceive the individual in this light as well. As this process begins, these needs will not be well-formulated in the minds of the individual or those that are attracted to the individual. Thus, the individual, or another that the individual is attracted to, symbolizes or embodies that which each subconsciously desires to become. Of course, if the individual or the other is in an existing primary relationship, this situation promotes its own kind of conflict and confrontation.

The choices that are made regarding what to do about these attractions are critically important from a karmic point of view. When this situation is encountered, I would recommend the following strategies:

1. If the individual, or another who is attracted to the individual, senses that the existing relationship has simply run its required course, that there is no more growth or usefulness in it for either partner, it is important to be totally open and honest about the new attraction. The existing relationship should be ended before involvement with another begins. Partners in the old relationship should spend whatever time is necessary to resolve the issues or conditions that led to the parting; to reflect together upon the entire nature and experience of the relationship so that it can be concluded in a positive way. Reflecting together can bring maximum knowledge to both partners about the nature of themselves and the relationship, and also can promote an ending with a minimum of hard feelings. The point is that this process can symbolize the fulfillment of an existing relationship that has run its natural course for this life.

2. If the individual, or the other who is attracted to the individual, senses limitation in the existing relationship because of the way that it has been defined, and yet also senses that it can change, then the person should spend the necessary amount of time working toward that end. New attractions to another, or of another toward the individual, simply represent a "sign" to the person that something is amiss within the existing relationship. This "sign" should be interpreted as a need to confront whatever problems exist so that the necessary metamorphosis can occur. If time is spent working toward this end and has produced nominal or no change, then the individual is karmically free to move on. The resistance of the partner now determines the course of action for the individual. The resulting emotional blow to the partner can perhaps induce a subsequent growth process for that person.

In general, the key to working positively with this process is to be as open as possible to the new needs of the partner, and for the partner to be as open to the new needs of the individual. It is a time for change and growth and will require that each person listen to the other in order to identify reality as it exists for each. In this way, each can give to the other what is needed according to their reality. If the partner or the individual experiences restrictions of growth in the relationship then confrontations can occur and the relationship may end if one or both of the individuals are defensive or closed to the new needs of the other.

Physical problems that can arise during this evolutionary process will manifest themselves in the lower back, kidneys, spine, toxicity of the blood, intense headaches, hearing problems, urinary tract infections, congestion in the liver and problems related to the heart chakra.

PLUTO THROUGH THE EIGHTH HOUSE

The archetype or theme during this evolutionary process is one wherein the individual will confront all areas or dynamics within him or herself that are stagnated and non-growth oriented. Internally, these areas or dynamics can be of an emotional, intellectual, physical or spiritual nature. Externally, any area in the individual's life can be experienced in this way — career, relationships, family, money and possession issues, how the life has been externally structured, positioned and defined, and so forth.

The ways in which the individual has defined and related to any or all of these inner and outer dynamics must be fundamentally reformulated to reflect the need for growth. Any or all of these dynamics can be experienced as limiting further growth. A loss of meaning and significance given to these areas and dynamics is a common experience during this time and creates internal and external confrontations. Such confrontations reflect the need to metamorphose and redefine the areas that are limiting further growth. Commonly, however, the areas that need to be redefined are the ones that the individual has overly invested him or herself in. In other words, the individual has put his or her life together in certain ways with respect to values, ideas, beliefs, and desires — the "bottom lines" of the reality orientation. A sense of limitation and stagnation occurs because the individual now senses that there are different ways to define and relate to these dynamics or areas. To change and redefine the "bottom lines" is to risk fundamental insecurity because of the level of importance and focus that the individual has given to them.

Thus, the choices that the individual must make during this process are very important. The choices will determine how this evolutionary necessity is experienced. Positively, the individual will acknowledge what must be confronted and changed. Accordingly, he or she will implement the necessary strategies to effect the transformation. New emotional responses, ideas, directional shifts in life, value associations and so forth will be put in motion with faith. The knowledge of why or where these changes will lead may not be known prior to the initiation of the change itself. Commonly, this knowledge will occur after the changes are implemented. Faith is required. The key is to move forward with the new impulses, feelings, ideas and desires that will originate in the Soul. Like a volcano erupting, these new impulses will rise into the individual's conscious awareness from unconscious depths. The compression of internal energy that this process promotes forces these new seeds into consciousness.

Messages from the external environment via those close to the individual can also aid in understanding what must be done. These messages can be of a positive or negative nature. Positively, others will promote and encourage the growth needs of the individual. Negatively, others will create confrontations and attempt to hold the individual back because of the threat to their own security if the person changed. The individual may be required to eliminate from his or her life those who are attempting to hold him or her back. Similarly, the individual must allow and encourage the growth needs of others with faith and courage. For those that are close to the individual, this may create fear and insecurity because of the way that the relationship had been defined up to this point in time. Yet, the individual must approach this necessity for change in those close to him or her with faith. If this process can be approached in this way, the potential for a complete regeneration through the elimination of outmoded and growth-inhibiting dynamics and existing orientations to his or her life will occur in a positive and noncataclysmic way.

The potential for cataclysmic events will exist for those who resist this evolutionary necessity. They may experience the enforced removal of the dynamics or areas in which they are overly involved to promote the required changes. This period can be one of the times in life when the changes enforce accelerated development of the evolutionary intent as seen in the main karmic/evolutionary dynamic in each birthchart. The magnitude and quality of the cataclysm is relative to the individual's level of resistance and the capacity to understand and accept what must be done. The negative choice of resistance promotes a loss of perspective as these enforced changes take place. The inner sense of personal implosion and stagnation can simply consume the person. He or she can feel as though they are being dragged into a bottomless vortex where no light can penetrate or escape. The forces of the individual's unconscious realms become stronger than the egocentric ability to resist. As a result, the individual can experience emotional winds, moods, thoughts and impulses that do not seem to originate from his or her own con-sciousness, leading to a sense of powerlessness and confusion.

From an evolutionary perspective, this time frame correlates to a purging of all the old behavioral associations and patterns rooted in the unconscious that have been dictating the individual's respon-ses to his or her life. The purging of these patterns and associations will allow for a progressive regeneration and rebirth of the individual as new patterns and associations replace the old. This purging can include the loss of external situations or conditions — a career, a relationship, friends, money or possessions, and so forth. The exter-nal purging reflects the inner evolutionary need.

From an experiential perspective, this process is intense. The individual can alternate in a compulsive and apparently uncontrol-lable way between moments of intense highs and intense gloom or dark moods. The highs occur through a sudden glimpse of the light of understanding. The lows occur through the sense that all is lost, meaningless, and out of control. As the ego is buffeted this way and that by the inner storm of clashing emotional, intellectual, spiritual and physical states, the individual can feel as though he or she is on the brink of death. Again, resistance to the necessary evolutionary changes is the cause of these effects.

This evolutionary process promotes an intense karmic time. Prior-life actions, as well as actions in this life, that have created negative or difficult karma can catch up with the individual at this time. The effects and confrontations make the individual aware of his or her motivations and desires. The opportunity to resolve old issues and karma is at hand. The purging of old karma must take place so that the individual can be free to move forward upon his or her life journey. Positively, karmic fruition can occur relative to prior-life actions as well as actions in this life. The goals and desires that the individual has had may now culminate in actualization and fulfill-ment. The positive culmination itself promotes the need for new directions as that which had been desired comes to completion.

The purging that this evolutionary process induces can only lead to a regenerated and reborn individual who will emerge with new desires, ideas, emotional responses, and understanding as to the nature and purpose of his or her life. The process is similar to going off to war and experiencing its devastation, misery and pain, and then returning home with the battle scars in place. Similarly, the individual uses the knowledge and perspectives gained from this evolutionary process to motivate him or herself to move forward with life, resolving not to repeat the mistakes of the past.

Physical problems that can occur during this process can manifest in the pancreas, colon and intestine, as latent genetic disorders, a mutation of cells leading to cancer, or imbalance in the white and red blood cells. A rise of kundalini energy may create severe physiological and neurological disorders if the energy is not properly channeled. Spinal problems may develop, as well as temporary paralysis of the legs or arms, spontaneous out-of-body experiences, and problems associated with the coccyx or naval chakra.

In summary, this evolutionary process promotes a necessary death of all old and outmoded patterns of emotional, intellectual, physical, or spiritual behavior and orientation. It releases and purges the cumulative negative karma from this and other lives, or brings to culmination positive karma based on previous goals and desires. It removes all internal and external barriers with which the individual is overly identified that prevent further growth. Out of the ashes of this necessary destruction will arise a regenerated individual who is free to continue on his or her evolutionary journey.

PLUTO THROUGH THE NINTH HOUSE

The initiation of a general growth pattern will be the theme during this evolutionary process. As a result the individual will need to confront the limitations in his or her existing philosophical principles and convictions. In order for the growth to occur, the individual must allow a sense of intuitive guidance to lead to the initiation of the appropriate strategies. By inwardly centering upon the intuitive guidance, the individual can experience "visions" about the nature of his or her future. New goals, ideas and directions can be implemented that are based on these visions or intuitive promptings. The individual must learn to trust these intuitive impulses even if they tend to threaten existing reality orientations.

This evolutionary time frame is generally very positive in nature. Generally, the individual will be very future-oriented and full of optimism with respect to future plans. It is a time for setting new goals, and for implementing new plans or directional shifts that reflect these goals. Yet the need to confront existing reality in order to accommodate the new plans, goals, and future-oriented intuitive promptings can be difficult for some. One's philosophical orientation and beliefs which have shaped the directions and goals of one's life up until this point may conflict with the new desires and needs. If so, the individual will draw circumstances that directly challenge and confront those existing orientations. The outer confrontation will reflect the inner need to confront these limitations. However, the tone and nature of these external confrontations will tend to be very positive. Other people can now confront and argue with the person in such a way as to point out the weak links or limitations in his or her present orientation. Their motives will be positive and sincere as they encourage the individual to move on with life. Opportunities to expand can come from the external environment as well, and will be reflected in the main/karmic evolutionary dynamic in the birthchart, the experiences associated with the sign of the Ninth House cusp that Pluto transits, the natal placement of Jupiter if transited by Pluto, Jupiter progressed to Pluto, or Pluto's overall condition in the solar return chart when it is in the Ninth House.

This evolutionary time will deepen the individual's intuitive faculty. The progressive ability to understand the nature of his or her reality, past, present and future, will manifest itself intuitively. This understanding will not be a product of deductive analysis. Intuitive insights and knowledge simply emerge into the consciousness. The individual will desire to move forward now, and the past can be perceived as a restraint that prevents growth. Restlessness will create a sense of dissatisfaction with respect to the present reality. The individual will know why he or she must make the required changes prior to the change itself. He or she will know the meaning of the potential opportunities and new strategies as they unfold. Even the most stubborn and resistant individual will understand this on an intuitive basis.

For those who are resistant by nature, who attempt to hold onto the past, the changes occur anyway. For some these changes can be hard, yet they will know why they must experience them. If the degree of resistance requires the forced removal of certain external conditions in order to permit inner change, the famous silver lining in the cloud will exist. In hindsight the individual will see the wisdom and positive development that the removal promoted. Again, on a bottom line basis, the individual is truly desiring to move and change anyway.

It is a fairly common experience for many people to experience the desire to be completely free from all existing obligations and reality conditions in their life during this time because of the deep inner urge to grow in an unrestrained manner. Yet, existing reality must accommodate, guide and shape these growth needs in order to control the growth. The danger is to expand in uncontrolled and unfocused ways, which will lead to disorganization or moving too fast. Positive growth can only occur through referring the new growth areas to the existing reality references in order to assimilate the growth. By throwing off the past entirely, no assimilation of the new growth patterns can occur. Without an existing bottom line to refer the new growth urges to, the individual can become lost and chase one new direction or impulse after another without reason. Only the aspects of the past or of any dynamic that prevents growth should be thrown off.

As a result of this evolutionary time, the individual's philosophical or religious convictions and principles will be expanded upon and transformed. These principles have served as the basis of how the individual has been interpreting and understanding his or her life specifically, and life in general. The individual's understanding of life based on his or her principles and convictions promotes conceptions of what life is and how it should be lived. Inwardly the individual will sense the limitations in his or her philosophical concept of reality, and outwardly draw the required circumstances to point out these limitations. Again, the recommended strategy is simply to allow the deep inner intuitive sense of simultaneous limitation and expansion to guide the individual to new realizations that will allow for transformation of the old. Outwardly, the strategy is to listen to the ideas and thoughts of others in a responsive way: to take in the information and to consider its merits rather than defensively rejecting it in a reactionary manner. If the new thoughts and ideas feel intuitively right, then the individual should incorporate them into his or her existing philosophy. If this requires changing, redefining or eliminating preexisting ideas or principles, then they should do so.

Physical problems that can manifest themselves during this time can occur in the pancreas, thyroid, pituitary gland, liver, sciatic nerve, and the sacral chakra.

In summary, this process promotes a cycle of intensified growth and general optimism. It deepens the intuitive faculty and one's trust in intuition. Opportunities for growth will be manifested in the areas that involve the individual's evolutionary and karmic needs and necessities.

PLUTO THROUGH THE TENTH HOUSE

A new chapter in the individual's life will characterize the theme or archetype of this process as an old cycle or phase is coming to a close. As this process begins, the individual must engage upon an active self-reflection process that examines all the internal dynamics that have created the outer reality structures of life. Through this self-reflection, self-knowledge can occur with respect to why the person has put his or her life together in the ways that it has been. This knowledge can be used as the basis for eliminating outmoded and crystallized emotional, intellectual, physical and spiritual patterns of all kinds. That which must be changed can lead into a new cycle of development.

The individual must learn to accept and acknowledge the responsibility for his or her own actions, and resist the temptation to blame others, or to throw off the past wholesale. New patterns of growth and desires must evolve, yet must be structured into the context of existing obligations and responsibilities established before this process began. In addition, this evolutionary time period promotes the need to become aware of the basis of the individual's judgments and how those judgments reflect a standard of conduct that may no longer apply. Old attitudes of all kinds must be reflected upon to see if they still apply, or if they are outmoded and preventing further growth.

Common psychological symptoms during this time are weariness leading to a sense of emptiness and personal meaninglessness. This weariness is not only connected to this life, but is also based on many other lives that have been structured upon the old and now crystallized emotional, intellectual, physical and spiritual patterns. This weariness is necessary for it promotes reflection upon what needs to be changed. The sense of meaninglessness associated with current circumstances and the overall nature of the individual's reality is also necessary because it encourages a deep Soul-searching that revolves around the need to restructure one's life to induce new meaning, purpose and relevancy.

Self-reflection promotes awareness of past modes of operation at all levels and induces the phenomenon of involution — the destruction of forms and structures. Through redefinition or elimination of all the old forms and structures, an evolution takes place: new forms and structures incorporating new ideas, opinions, beliefs, emotional and physical responses, attitudes and equations of all kinds. These forms and structures determine the totality of the individual's inner and outer reality and how he or she has viewed and integrated him or herself into the society in which they live.

Because the Tenth House, Capricorn, and Saturn all correlate to our need to establish and integrate our own authority within the context of societal authority (career) this evolutionary process promotes a time wherein the individual must examine the nature and basis of his or her existing career. This must occur to determine whether it still contains the potential for meaningful self-expression. If not, then this time demands that the individual begin the process of creating changes — a new career, or redefinition of an existing career. For some this evolutionary necessity will translate into the removal of a career. For others, the level of frustration, futility, or nonrelatedness simply intensifies with respect to the existing career. For still others, problems with a supervisor or with co-workers suddenly begin to arise. For some, a promotion or an increase of responsibility can occur even as they are feeling less than satisfied in their career. This situation creates an evolutionary test for these people that revolves around the issue of what they do for practical reasons because of the promotion or increased recognition, versus the courage to move on to reflect their own growth needs. For those who are feeling good about their career, a promotion, increase in responsibility or higher visibility will create the new meaning that they were seeking. For those who either are not in a meaningful career, or who have been prevented from actualizing or fulfilling their career needs, this is a time to change in order to find or create a new career. In so doing the person will find the sense of personal meaning that he or she is seeking.

During this evolutionary process it is also time to examine the basis of personal judgments and the standards of conduct from which these judgments originate. These standards have evolved through parental and societal conditioning, personal actions that have led to many experiences through which a judgment was made, and prior-life conditioning influences that are linked to these two dynamics. Judgments equal mental and emotional attitudes and responses based on these conditioning dynamics. The individual must reflect upon his or her judgment patterns for two reasons: 1. to determine their impact on his or her own and others' actions and projected actions in the future, and 2. to determine if these judgments are negatively influencing further growth opportunities for the individual and others. The key is determining the negative impact of these judgments is to see if they induce a blocking action as the individual contemplates future possibilities, and to reflect upon why he or she can draw negative judgmental reactions at this time. If the individual detects a blocking effect from within him or herself, or a blocking effect in the form of negative feedback from others or society, as he or she contemplates future possibilities, then two situations can exist: 1. the blocking activity reflects a resistance to moving forward because of fear or old judgmental patterns that no longer apply, or 2. the old attitudes, associations, or future possibilities are not relevant.

In the first condition the individual must examine the basis of the resistance, fear, or old judgmental patterns to be free of them so that the he or she can move forward with his or her life. In the second condition the individual must be open to the environmental signs that create this reaction. In other words, old attitudes and associations based on existing judgments will be met by negative feedback from others. Thus the individual must be willing to accept this and change as necessary. With respect to future possibilities that draw negative reactions from the environment, the individual must determine if these possibilities are actually "real" and legitimate as linked with the ongoing context of his or her life, or if they are actually pie-in-the-sky scenarios that are not possible within the life context. The environmental signs will help the individual determine this because of the reflection that it creates within the individual. Sometimes these environmental signs will be of a positive and negative nature to further help the individual implement the proper course of action.

If a course of action does seem right, then the individual must put him or herself in motion in determined and persevering way. The environmental signs will be positive and the gates will be open. Certain conditions may have to be redefined or eliminated in the individual's life in order to reflect these new directions that promote meaning in his or her life. If these directions or possibilities are wrong then all doors will be blocked no matter what the individual does.

An example of this process is reflected in a client who came to see me when she was fifty-six years old. Her children had grown up and she had not prepared to do anything else with her life. Pluto was transiting her Tenth House cusp at this time. Her birthchart showed the natural signature of a psychologist. I suggested that she go to school and get her credentials. She initially rejected this idea because of existing societal standards and judgments about age. "I'm too old", she said. In addition, her husband wanted her to hang around the house and to continue catering to his whims and needs. His argument was the same — "You are too old." In addition, she had been conditioned to be a "Betty Crocker" type by her parents. I told her she had nothing to lose and everything to gain by at least looking into further schooling.

On the one hand, I manifested a negative environmental sign because of the feedback I gave with respect to her old judgmental attitude and associations about age. On the other hand, I manifested positive feedback with respect to what she could actually do and in fact was thinking about deep within herself. Well, she did. The school she checked out was more encouraging and offered her financial assistance. She had to terminate her relationship with her husband because he simply did not want her to do this. He felt threatened and insecure. She tried to fit this new direction in her life within her existing responsibilities and obligations — her relationship with her husband. Yet, because he was emotionally unable to accept this new direction in her life, she had to end the marriage in order for her own growth to occur. This one negative environmental "sign" linked with my negative feedback about her old judgmental attitudes made her reflect upon the fact that both the school and her astrologer manifested positive signs. She obtained her Master's degee in psychology, began a practice for senior citizens in which she tried to strip away their assumptions about age, and then went on to get her Ph.D.

Physical problems that can manifest themselves at this time can occur in the bone marrow, skeletal system, pituitary gland, or immune system. Growths of all kinds can occur on the skin or scalp, enzyme imbalances, acceleration of the aging process or, conversely, rejuvenation of personal vitality and a slowdown of aging, degeneration or regeneration of organ tissue, or problems associated with the coccyx chakra.

In summary, this process demands a total redefinition of all the outmoded structures of the individual's life. A new phase or cycle is beginning. Self-determination, perseverance, and courage to redefine the orientation and directions of one's life must occur.

PLUTO THROUGH THE ELEVENTH HOUSE

The need to sever all attachments that bind the individual from further growth will characterize the theme or archetype during this evolutionary process. This process will rapidly accelerate the growth of the individual. This acceleration will promote the sense of being cramped by the existing reality conditions in the person's life. The sense of being cramped creates an internal compression of the individual's consciousness. This compression cracks open the individuated unconscious wherein lie all the issues and dynamics that the person has heretofore suppressed, including the memories of this and other lives in complete detail, and information concerning the future. This information impacts on the conscious awareness so that the individual can be free from the past and present that bind him or her with respect to future growth needs. The intense need to be free from the past and present creates cycles of emotional, intellectual, physical and spiritual detachment from existing conditions in the individual's life. In these cycles of detachment the individual can objectively survey all the inner and outer dynamics that have led to this moment in time. This objective survey allows for a dispassionate analysis of what needs to be changed or eliminated in order for the new evolutionary impulses to proceed. These cycles of detachment clash with cycles of resistance to the necessary changes as the individual can feel threatened and insecure about where the future-oriented thoughts and impulses will lead.

Many individuals will feel that they have no control over these new impulses and thoughts—they seem to have a life of their own and simply arrive in the conscious awareness of their own volitions. A key to positively dealing with these future-oriented thoughts is to monitor the element of repetition in these thoughts. In other words, the person should move toward actualizing thoughts that keep repeating themselves, not those that appear one or two times. The thoughts and impulses that appear once or twice are either "wishful thinking" or ahead of their time—they may be relevant many years down the road but not now. In addition, the individual should monitor external messages from others that are repetitious by nature concerning the individual's future. These external signposts will help the individual validate what he or she is sensing and experiencing within.

Because there is a need to alter or eliminate inner and outer conditions, situations or dynamics concerning the past, the individual will necessarily have to confront them in order to be free. Externally, this process can manifest itself as sudden clashes with friends or associates. Some of these clashes or confrontations will promote a surfacing of a residual karma that has not been resolved between the individual and others. Thus the opportunity to resolve and be freed from this karma is at hand so that the individual can move forward. The sudden confrontations or clashes can be quite confusing for the individual and those with whom he or she clashes. What had heretofore been a solid, close and trusted association suddenly erupts into clashes and confrontations that surprise both people. A lack of perspective or understanding as to why these confrontations are occurring, and a possible termination of the relationship because of them is very common at this time. The recommended strategy to employ now is to terminate the attachments to others who are attempting to hold the individual back from further growth, to resolve whatever karmic issues have created the confrontation, and to let certain relationships go if a natural parting of the ways is sensed. The strategy is to nurture relationships with those who encourage the necessary growth needs of the individual, who help him or her to objectively understand the nature of his or her future-oriented thoughts and impulses. The necessary termination of certain existing relationships will allow for new people, associations, or close relationships to come into the person's life who will naturally resonate with his or her new needs and desires at this time.

The need to sever all attachments to existing ideas and mental orientations and emotional, physical, or spiritual responses is necessary now. New patterns and orientations at all levels are now in the process of rapidly evolving into the individual's consciousness. An acceleration of the evolutionary pace can promote "evolutionary leaps" with respect to the natural condition in which the individual entered this life. Thus, the need for new friends, associations, or group activities is encouraged. The need to sever attachments to existing ideas, emotional responses and so forth will make the individual feel different about his or her life at this time. Others who are on the same path, or who have evolved or are evolving to where the individual is now headed, can help reconcile the individual to what is happening to him or her. This kind of objective validation will help the person take charge of his or her life, and to become free of the past that prevents them from necessary growth.

It is also common during this process for the individual to draw others to him or her who are experiencing these same needs. Unsettling and restless vibrations will seem to permeate the individual's inner and outer atmosphere. The individual should encourage the growth needs for all those with whom he or she is in close association at this time even if it means a change in form or termination of a relationship. Approached in this way, positive karma will be maintained or established. If the individual attempts to restrain the growth needs of other people, then the relationship potentially will terminate in a difficult and negative way. Terminations of this kind breed negative karma or unresolved issues that will have to be resolved at another time.

Physical problems that can manifest themselves during this process occur in the parasympathetic and sympathetic nervous systems as varying levels of stress, problems associated with any genetically weak areas that can be triggered because of the stress, neurological problems that may lead to a stroke, problems with the lungs or breathing, spinal cord, rashes on the skin, headaches from tension in the medulla chakra (base of the skull), dizziness, disturbances in the rhythm of the heartbeat translating into cycles of high and low blood pressure, and problems associated with the coccyx and cervical (throat) chakras.

In summary, this process accelerates the evolutionary pace of the individual. The need to be liberated from the past so that rapid growth can occur will promote new and potentially radical thoughts and impulses upon the screen of the individual's consciousness. The individual must terminate that which binds him or her to the past, and follow the future-oriented internal and external messages of a repetitious nature. In this way, the person can make large-scale evolutionary advances.

PLUTO THROUGH THE TWELFTH HOUSE
The need to become aware of the nature of personal dreams,

illusions and delusions will characterize the theme or archetype of this evolutionary process. Thus, the need to deal with and acknowledge reality as it is will result. Commonly, the dreams and wishes that the individual has or has had will be manifested in reality at this time. In so doing, the reality of those dreams and wishes will be experienced as something less than what the individual thought they would be. These dreams and wishes will be linked with desires of a separating nature. By desiring and willing these dreams into reality, the individual is being taught the spiritual lesson that he or she is a co-creator of reality, a co-creator linked with the Source that has ultimately created everything. By experiencing the reality of these dreams and wishes as something less than what the individual thought they would be, he or she will sense that there is something more to life—a sense of something missing. This feeling leads to the sense of ultimate alone-ness at a gut level that cannot be quenched by any external or separating desire, nor fulfilled by the existing conditions of the per-son's life. These separating desires can bring a sense of temporary satisfaction, but it is soon replaced with the knowledge that there is more. Accordingly this is an excellent time for the individual to understand and get in touch with the basis of his or her separating desires.

The need to merge with the source who created the individual will promote real cycles of confusion, alienation, dissassociation and meaninglessness from the individual's existing reality at this time. Thus, this can be a time in which the individual experiences an essential and necessary disillusioning process. The current reality coupled with one dream/illusion after another promotes the sense of standing upon a precipice through which the light of the world (the past) is contrasted with the darkness of infinity, the absolute and the universal that lies before the individual. The need to merge with the Source, the Universal, requires that the individual leap into the abyss with faith so that all barriers preventing this conscious union can be dissolved. Yet this evolutionary need is commonly met with resistance and fear. Just as the wave can resist returning to the sea for a time, so too does the egocentric identification within the individual resist surrendering to the Cosmic Ocean or Being from which he or she originated. The resistance causes the cycles of alienation and dis-association from the individual's existing reality. The reality is now permeated with a sense of meaninglessness. Externally, certain areas the individual has overly identified with can be removed or dissolved, and he or she will seem to have no power to prevent this. This purg-ing and elimination must occur to free the person to move forward, and to teach him or her that there are larger forces within the universe that oversee and guide the development of all aspects of Creation, individually and collectively.

Because of this dissolving process, the individual will commonly experience intense dream activity at this time. The dream activity promotes a release of unconscious content so that a psychic purging can occur that allows for a new cycle of growth and direct connection to the Source to occur. These dreams will be of three kinds:

1. past life dreams which the individual must look at in order to see their connection to current conditions in his or her life. These past-life dreams will be focused upon one particular lifetime, or many lifetimes lumped together in one dream that is highly symbolic in nature.
2. dreams that are based on some current experience or circumstances in the individual's life that provide deeper meaning as to the reality or nature of that experience or circumstance.
3. "superconscious" dreams in which the individual finds him or herself in astral or causal realms in order to receive some kind of instruction, knowledge, or revelation as to the nature of his or her existing reality, or that is linked to some question that has been posed by the individual. Upon waking the individual will have two experiences with respect to these types of dreams: either total and vivid recall in such a way that wherever they went seems more real than the immediate "reality" of their current life conditions or no recall, yet the knowledge that they had been elsewhere.

The strategy recommended at this time is to keep a dream journal so that a record can be made of these dreams. This is necessary because dreams follow the Moon's twenty-eight day cycle. Thus, the context of one's dreams during this natural cycle will allow for any one dream within the twenty-eight day cycle to be interpreted accurately versus trying to understand the meaning of any one dream without the context of a cycle of dreams to refer to.

The pineal gland is naturally stimulated during this evolutionary time frame. Thus, the natural transcendent biological chemical melatonin is secreted into the individual's body. This promotes altered states of consciousness that induce hypersensitivity throughout the organism physiologically and psychologically. The person is wide open to the impact of internal and external vibrations of a positive and negative nature at this time. Negatively, this promotes receptivity to thoughts and vibrations of a delusive nature leading to cycles of confusion, alienation, disassociation and so forth. These thoughts and vibrations can emanate from within the individual, or they can emanate from the thoughts of others received via sensitivity to the collective consciousness.

Positively, this experience promotes receptivity to all thoughts and vibrations of a spiritual or transcendant nature leading to divine illumination, revelation and communion. As a result, the individual should align him or herself with a transcendant belief system because this will allow for the necessary perspectives and knowledge that will promote a conscious development of the spiritual component, and allow for the development of techniques that facilitate the awareness of what is truth versus what is illusion. If the individual does this, then there will be no barriers preventing a conscious relationship with the Source, and the individual will understand the basis of his or her fears that promoted a resistance to this necessary merging. In addition, the individual will understand the nature of his or her desires, and be free to live the life that he or she is destined to live (from an evolutionary and karmic point of view). This freedom will replace being enslaved by his or her unending desires that create one dream and illusion after another.

The individual is also learning to live in the moment at this time, the eternal now. Living each moment as it comes will allow the individual to evolve into his or her future as it is unfolded in each successive moment. In this way, the person can take stock of who he or she is at whatever moment: reality as it exists at each moment in time. This process will promote clarity of self-understanding and a clear perspective about the past that has led to the present. In addition, it will allow the individual to sense the divine guidance that is moving through him or her in such a way as to reveal not only the individual's real nature, but to also reveal the future path as it unfolds in each moment.

During this time the individual must also balance necessary "down" time (to be alone without activity) with external time and activity. The down and external times, when responded to according to natural flow or rhythm, will promote a continual state of clarity. This rhythm will fluctuate and be unpredictable. If this rhythm is not followed, then loss of perspective will occur with the resulting confusion and alienation.

Physical problems that can manifest themselves at this time will occur in the endoctrine system, immune system, hypothalmus, and pineal gland. The person may relive old ailments because the organism is trying to cleanse itself. There may be supersensitivity to impurities in food or drink, transitory allergic reactions, intensification of existing or latent genetic problems, enzyme imbalances, problems associated with the pancreas, liver, colon, intestine, problems that cannot be diagnosed by an allopathic physician, spontaneous out-of-body experiences, and problems associated with the sacral or crown charkras (a burning in the top of the head or the middle of the abdomen). Lots of pure fluids, hot baths with epsom salts, relaxation exercises, meditation, and yoga are good now.

In summary, this evolutionary time demands that the individual merge the self with a transcendental belief system to promote a conscious awareness of his or her natural spiritual identity. In so doing, the knowledge of being a co-creator with the Source can occur. The cycles of dissolution, confusion and alienation from existing reality serve as the "signs" that will show the individual where the Ultimate Meaning that he or she is seeking lies. The nature of personal desires, dreams and illusions can now be understood. Accordingly, the individual now has the opportunity to consciously align his or her personal actions, desires, and identity with the Source who is attempting to guide the person upon his or her evolutionary and karmic path. In this way, the individual can walk hand in hand with the Source so that his or her personal actions and desires are in harmony with the will and desire of the Source as It expresses Itself through the life plan of the individual.

CONCLUSION

In this first volume of *Pluto: The Evolutionary Journey Of The Soul* the focus has been to present an understanding of what the Soul is, and what the inner dynamics are that lead to its progressive evolution toward a perfected state of self-realization.

Within this general understanding, the specific evolutionary lessons of any individual, in any one life, have been explained from an astrological point of view. The specific nature of one's personality, and the orientation to phenomenal reality that the Soul-generated personality creates, describes not only how we experience life, but also the specific circumstances and parameters of what a particular life is about. It is hoped by the author that this way of understanding our lives will help in promoting not only a cooperation with the evolutionary intent of the life, but also an acceptance of the basic fact that we are responsible for what we have created from our prior actions which are rooted in the dual desire-nature of our Souls.

This book has been aimed at the person who asks the questions "Why am I here, and what are my lessons?" These questions are becoming more frequently asked as the collective insecurity within all Souls intensifies due to Neptune's passing through Capricorn, and Pluto's return to its home sign Scorpio. For all of you who have asked these questions, the author hopes this book has aided you in your answers. For the professional astrological counselor whose clients ask these questions, it is hoped that this book will facilitate your ability to guide them into their own understanding of these questions.

The more people who attempt to understand their lives in this way, and thus cooperate with the lessons at hand, the more the potential and probability of collective and individual cataclysmic evolutionary events is decreased. Again, it is the collective and individual resistance to the evolutionary necessities at hand that create the very real phenomena of cataclysmic events leading to the required evolutionary progression. It is in this spirit that the author has written this book.

In the second volume of *Pluto: The Evolutionary Journey of The Soul*, which will follow at some point, Pluto's role in our relationships to others will be covered. The issue of prior-life connections, the reasons or karma that has brought us together in this life, and the lessons that we are working on together will be explored in depth through synastry comparisons, as well as how these themes are reflected in composite charts. The second volume will also discuss Pluto's correlations to anatomy and physiology, and will present a comprehensive explanation of the chakra system. In addition, Pluto's role through history will be examined and linked to the choices humanity faces over the next few decades that will determine the collective reality that we will all live within.

After reading this first volume, I would invite you to write to me if you have questions about this material. If you would like your own birthchart analyzed with respect to your evolutionary and karmic necessities, I would be happy to do this for you on tape. In either case you can reach me through Llewellyn Publications (the information on how to do this is in the front of the book). Send me a SASE and I will send you all the information that I will need from you, plus cost, etc. If you would like me to do a workshop or lecture in your area, you can also reach me through Llewellyn.

I sincerely hope that you have gained new understanding from this book, and that you can apply this understanding to your own life and to those that you may work or be involved with. God Bless.

On the following pages you will find listed, with their current prices, some of the books now available on related subjects. Your book dealer stocks most of these and will stock new titles in the Llewellyn series as they become available. We urge your patronage.

TO GET A FREE CATALOG

You are invited to write for our bi-monthly news magazine/catalog, *Llewellyn's New Worlds of Mind and Spirit.* A sample copy is free, and it will continue coming to you at no cost as long as you are an active mail customer. Or you may subscribe for just $10 in the United States and Canada ($20 overseas, first class mail). Many bookstores also have *New Worlds* available to their customers. Ask for it.

In *New Worlds* you will find news and features about new books, tapes and services; announcements of meetings and seminars; helpful articles; author interviews and much more. Write to:

Llewellyn's New Worlds of Mind and Spirit
P.O. Box 64383-L296, St. Paul, MN 55164-0383, U.S.A.

TO ORDER BOOKS AND TAPES

If your book store does not carry the titles described on the following pages, you may order them directly from Llewellyn by sending the full price in U.S. funds, plus postage and handling (see below).

Credit card orders: VISA, MasterCard, American Express are accepted. Call toll-free in the USA and Canada at 1-800-THE-MOON.

Special Group Discount: Because there is a great deal of interest in group discussion and study of the subject matter of this book, we offer a 20% quantity discount to group leaders or agents. Our Special Quantity Price for a minimum order of five copies of *Pluto* is $51.80 cash-with-order. Include the postage and handling charges noted below.

Postage and Handling: Include $4 postage and handling for orders $15 and under; $5 for orders *over* $15. There are no postage and handling charges for orders over $100. Postage and handling rates are subject to change. We ship UPS whenever possible within the continental United States; delivery is guaranteed. Please provide your street address as UPS does not deliver to P.O. boxes. Orders shipped to Alaska, Hawaii, Canada, Mexico and Puerto Rico will be sent via first class mail. Allow 4-6 weeks for delivery. **International orders:** Airmail – add retail price of each book and $5 for each non-book item (audiotapes, etc.); Surface mail – add $1 per item.

Minnesota residents please add 7% sales tax.

Mail orders to:
Llewellyn Worldwide, PO Box 64383-L296, St. Paul, MN 55164-0383, USA
For customer service, call (612) 291-1970.

URANUS
Freedom From the Known
by Jeff Green

The archetypal correlations of the planet Uranus to human psychology and behavior to anatomy/physiology and the chakra system, and to metaphysical and cosmic laws are covered in *Uranus*.

This book evolved in style and tone from an intensive workshop held in Toronto. In reading *Uranus* you will discover how to naturally liberate yourself from all of your conditioning patterns, patterns that were determined by the "internal" and "external" environment. Every person has a natural way to actualize this liberation. This natural way is examined by use of the natal chart and from a developmental point of view.

The 48-year sociopolitical cycle of Uranus and Saturn is discussed extensively, as is the relationship between Uranus, Saturn and Neptune.
0-87542-297-7, 192 pgs., 5 ¼ x 8, softcover **$7.95**

HOW TO PERSONALIZE THE OUTER PLANETS
The Astrology of Uranus, Neptune & Pluto
Edited by Noel Tyl

Since their discoveries, the three outer planets have been symbols of the modern era. Representing great social change on a global scale, they also take us as individuals to higher levels of consciousness and new possibilities of experience. Taken as a group, as they are in *Personalizing the Outer Planets*, the potential exists to recognize *accelerated* development.

As never done before, the seven prominent astrologers in *Personalizing the Outer Planets* bring revolutionary forces down to earth in practical ways.
- Jeff Jawer: Learn how the outer planet discoveries rocked the world
- Noel Tyl: Project into the future with outer planet Solar Arcs
- Jeff Green: See how the outer planets are tied to personal trauma
- Jeff Jawer: Give perspective to inner spirit with planet symbolisms
- Jayj Jacobs: Explore relationships and sex through the outer planets
- Mary E. Shea: Make the right choices using outer planet transits
- Joanne Wickenburg: Realize your unconscious drives and urges through the outer planets
- Capel N. McCutcheon: Personalize the incredible archetypal significance of outer planet aspects

0-87542-389-2, 288 pgs., 6 x 9, illus., softcover **$12.00**

Simple Natal Chart

You discover where your pluto is in your chart with this chart printout programmed and designed by Matrix. It will give all of the major aspects, the nodes mid-points, and more. Give exact birth time, date, year, and location.
NATAL CHART, APSO3-119 **$5.00**

To read about other Llewellyn Astrological Services, write for our FREE catalog.

Prices subject to change without notice.